HOW TO SOLVE
STATISTICAL PROBLEMS
WITH YOUR
POCKET CALCULATOR

To Dick

HOW TO SOLVE STATISTICAL PROBLEMS WITH YOUR POCKET CALCULATOR

BY VICKI F. SHARP

EAST CHICAGO PUBLIC LIBRARY
EAST CHICAGO, INDIANA

TAB BOOKS Inc.
BLUE RIDGE SUMMIT, PA. 17214

Appendix Tables A, I, and L adapted from Fisher & Yates: *Statistical Tables for Biological, Agricultural and Medical Research*, published by Longman Group Ltd. London (previously published by Oliver & Boyd Ltd. Edinburgh) and used with permission of the authors and publishers.

Appendix Tables B, C, D, F, and G adapted or reprinted with permission from *Basic Statistical Tables*, 1971, Copyright © The Chemical Rubber Co., CRC Press, Inc.

Appendix Table E reprinted with permission from *Nonparametric Statistics for the Behavioral Sciences* by Sidney Siegel, Copyright © 1956, McGraw Hill.

Appendix Table H adapted from Milton Friedman, "The Use of Ranks to Avoid the Assumption of Normality Implicit in the Analysis of Variance," *Journal of the American Statistical Association*, Vol. 32, p. 688, 1937.

Appendix Tables J and K reprinted with permission from *Elementary Statistics*, Fourth Edition, by Paul G. Hoel, Copyright © 1976, John Wiley & Sons, Inc.

FIRST EDITION

FIRST PRINTING

Copyright © 1982 by TAB BOOKS Inc.

Printed in the United States of America

Reproduction or publication of the content in any manner, without express permission of the publisher, is prohibited. No liability is assumed with respect to the use of the information herein.

Library of Congress Cataloging in Publication Data

Sharp, Vicki F.
    How to solve statistical problems with your pocket calculator.

    Includes index.
    1. Statistics—Data processing. 2. Calculating—Machine. I. Title.
QA276.4.S45   519.5'028'5   81-9245
ISBN 0-8306-0031-0   AACR2
ISBN 0-8306-1303-X (pbk.)

519.5
S531hP

Table Of Contents

Introduction vii

1 Math Fundamentals for Statistics 1
Introductory Test—Answers—Review of Arithmetic with the Calculator—Power of Numbers—The Memory Key—The Number Line—Arithmetic Operation Order—Reading a Statistical Formula—Test 1—Answers

2 Getting Acquainted with the Calculator 19
Mean—Standard Deviation—Standard Error of the Mean—Test 2—Answers

3 Review of Research Terms 27
Descriptive Statistics—Inferential Statistics—Population—Sampling—Experimental Study—Independent Variable and Dependent Variable—Null Hypothesis—Alternative Hypothesis—Level of Significance or Alpha—Sampling Distribution—Two-Tail Tests—One-Tail Test—Degrees of Freedom—Parametric Tests and Nonparametric Tests—Test 3—Answers

4 How to Choose Quickly a Statistical Test 34
Level of Measurement—Number of Groups—Nature of Groups—Test 4—Answers

5 Nominal Tests 41
Chi-Square (I) Test—Chi-Square (II) Test—McNemar Test—Cochran Q Test—Test 5—Answers

6 Ordinal Tests 65
Kolmogorov-Smirnov Test—Mann-Whitney U Test—Wilcoxon Signed-Ranks Test—Kruskal-Wallis Test—Friedman Test—Test 6—Answers

7 Interval Tests 100
T-Test (I)—T-Test (II)—T-Test (III)—One-Way Analysis of Variance, Randomized Groups—Scheffé Test—One-Way Analysis of Variance, Randomized Block Model—Test 7—Answers

8 Correlational Techniques   152
Contingency Coefficient—Spearman Rank Coefficient (rho)—Pearson Product-Moment Correlation—Simple Multiple Correlation—Test 8—Answers

9 Other Correlational Techniques   177
Point-Biserial Correlation—Kendall Coefficient of Concordance (W)—Kendall Rank Correlation (tau)—Test 9—Answers

10 Test Yourself   202

Appendix A   Statistical Tables   229

Appendix B   Statistical Formulas   250

Bibliography   254

Index   256

Introduction

With the aid of an inexpensive calculator, you can quickly solve a chi square or analysis of variance with no difficulty. You do not need a sophisticated math background to accomplish this feat. There are no math derivations to be learned. This is a self-instructional book that is easy to use and is activity-oriented. You can use this book as a self-contained package or as a supplement for any introductory course in statistics. It brings delight to many age groups. It can be used by gifted elementary school children, high school students, undergraduate and graduate, college students, and laymen. This volume can be rewarding to the teacher of statistics, test and measurements, or research. The material is presented in such an easy-to-follow manner that anyone can understand it and not fear the mathematical computations.

This book has some unique qualities. The ten chapters are independently presented. If you are acquainted with the fundamentals in one chapter you can skip it and proceed with another without losing content. The book can also serve as a long-range reference tool; it can be reused many times since it contains all the statistical tests that are commonly used. There is a rich assortment of practice problems with answers in each chapter. The numbers used in the problems are intentionally small to avoid laborious effort. There is a special chapter that provides you with problems which use all of the techniques covered in the book. There are also calculator shortcuts for solving the statistical problems. The book is made more entertaining and useful by a statistical map that shows you how to quickly choose a statistical test. Each statistical test is then explained step by step. I have also provided tables, formula lists, and suggested readings for your convenience.

Here is a brief synopsis of the book: Chapter 1 provides the necessary math for statistics. Chapter 2 presents exercises so you can use your own pocket calculator. Chapter 3 is devoted to a review of the research terms. Chapter 4 demonstrates a unique method of choosing a statistical test by using a specially designed map. This is a unique and useful tool that you can master in a minute's time. Chapter 5 presents a selection of nominal tests. Chapter 6 introduces ordinal tests. Chapter 7 explains interval tests. Chapter 8 is devoted to correlational techniques. Chapter 9 has other correlational techniques. Finally, Chapter 10 tests on the entire content of the book.

I would like to express my appreciation to the authors and publishers who gave permission to use the tables. I am particularly grateful to the literary Executor of the late Sir Ronald A. Fisher, F.R.S. to Dr. Frank Yates, F.R.S. and to Longman Group Ltd., London for permission to adapt tables III, IV and VII (Appendix tables A, I, and L) from their *Statistical Tables for Biological, Agricultural and Medical Research* (6th edition, 1974).

Many individuals have contributed to this book. A special thanks is given to editor Steve Mesner for his invaluable editing, and to Ellen Britsch who saw the value of this volume. Last, I wish to express my appreciation to Dick, my wonderful husband, for his support and helpful suggestions.

Chapter 1
Math Fundamentals for Statistics

Take the following test using your calculator. After you finish, check your answers. If you score 90 or better, it is not necessary to read this chapter; simply skip to Chapter 2.

INTRODUCTORY TEST

1. Solve the following decimal problems:

   (a) 6.83
        24.33
        30.22
       +4.22

   (b) 693.5
       − 18.2

   (c) 4.98
       ×3.9

   (d) $\dfrac{.0008}{.004} = $ _____

2. Round these numbers to the nearest hundredths place:
   - (a) 18.3567 _____
   - (b) 567.332 _____
   - (c) 82.899 _____
   - (d) 1.674 _____
   - (e) 1.943 _____
   - (f) 24.321 _____

3. Convert the following fractions to decimals:
   - (a) 1/2 _____
   - (b) 1/10 _____
   - (c) 1/9 _____
   - (d) 1/8 _____

4. Solve these problems:
   - (a) 986742.3 × 0 = _____
   - (b) 0/8 = _____
   - (c) 10 + 0 = _____
   - (d) 895 × 0 = _____

1

5. Solve the following problems:

   (a) $5^2 = $ _____
   (b) $\sqrt{100} = $ _____
   (c) $\sqrt{16} = $ _____
   (d) $8^4 = $ _____
   (e) $10^3 = $ _____
   (f) $\sqrt{25} = $ _____

6. Put the numbers 8, 9, 3, 2, 4 in memory plus then add memory plus to 88. What is your answer?

7. Put the numbers 9, 10, 3, 2, 1 in memory minus then add memory minus to 100. What is your answer?

8. Complete the following operations:

   (a) $(-5) + (-6) = $ _____
   (b) $18 - (-5) = $ _____
   (c) $(-6) \times 3 = $ _____
   (d) $(-9) + 6 + (-10) + 5 = $ _____
   (e) $18 - 3 = $ _____
   (f) $\dfrac{-30}{10} = $ _____
   (g) $(-8) \times 4 = $ _____
   (h) $\dfrac{-20}{-10} = $ _____

9. Solve the following problems:

   (a) $(6 \times 5) + 8 \div 4 = $ _____
   (b) $4^2 + 10 \div 5 + 6\sqrt{100} = $ _____
   (c) $9 + 3 + 9^3 - 6 = $ _____

10. Using the formula $s = \dfrac{1}{N} \sqrt{(N) \Sigma X^2 - (\Sigma X)^2}$
    find the standard deviation for the following raw scores: 1, 5, 8, 3, 2, 9, 15.

    s = _____

11. What do the following symbols mean?

    (a) ΣX _____
    (b) N _____
    (c) X _____
    (d) ΣX^2 _____
    (e) $(\Sigma X)^2$ _____

ANSWERS

Score 2 points for each problem that is done correctly with the exception of problem 10 which receives 8 points if answered correctly.

1.	(a)	65.6	(c)	19.422	
	(b)	675.3	(d)	.2	
2.	(a)	18.36	(d)	1.67	
	(b)	567.33	(e)	1.94	
	(c)	82.90	(f)	24.32	

3. (a) .5 (c) .1111111
   (b) .1 (d) .125
4. (a) 0   (c) 10
   (b) 0   (d) 0
5. (a) 25 (d) 4096
   (b) 10 (e) 1000
   (c) 4   (f) 5
6. 114
7. 75
8. (a) 11 (e) 15
   (b) 23 (f) −3
   (c) −18 (g) −32
   (d) −8 (h) 2
9. (a) 32
   (b) 78
   (c) 735

10. $s = \frac{1}{7}\sqrt{(7)409 - 1849} \qquad = \frac{1}{7}31.843366$

    $= \frac{1}{7}\sqrt{2863 - 1849} \qquad = 4.55$

    $= \frac{1}{7}\sqrt{1014}$

11. (a) Sum of the raw scores
    (b) Total number of scores
    (c) Raw Score
    (d) Sum of the squared scores
    (e) Sum of the scores squared

REVIEW OF ARITHMETIC WITH THE CALCULATOR

If you didn't do well on the preceding test, the following review will be helpful.

Arithmetic Symbols

Addition, subtraction, division, and multiplication are shown as follows.

How Shown	Represented by	Symbol	Expressed
Addition	plus	+	6 + 8
Subtraction	minus	−	9 − 4

How Shown	Represented by	Symbol	Expressed
Multiplication	dot	•	3•2
	parentheses	( )	9(3)
	times sign	×	3 × 6
	two symbols together	ΣX	ΣX
Division	horizontal bar (fraction)	—	$\frac{10}{5}$
	division sign	÷	20 ÷ 5
	traditional format	╱	50$\overline{)100}$
	slanted line	/	30/15

Decimals

Your calculator has a feature called a "floating decimal." Your decimal point key will place a decimal point in a number by pressing it at the proper time. For example, press the following keys:

Press	Display
4 • 1 5 × 6 =	24.9
6 • 2 4 + 3 • 4 2 =	9.66
• 0 8 • 0 2 =	4.00

Notice you do not need to move decimal places; the calculator does this mental work for you.

In this book you will change a fraction to decimal form by dividing the numerator by the denominator. Let us take the fraction ½ and change it to the decimal form. You divide 1 by 2. On the calculator you would push the following buttons.

Press	Display
1 ÷ 2 =	.5

The answer is .5.

Problem. Convert the following fractions to decimals: 1/8; 1/6; 1/5.

Answers

Press	Display
1 ÷ 8 =	.125
1 ÷ 6 =	.16666666
1 ÷ 5 =	.2

In rounding decimals you need to follow certain rules. To round any number, you first find the place in the number that has to be rounded. For example, if we are to round 8.4527 to the nearest hundredths place*, the place in the number that has to be rounded is between the 5 and the 2. You rewrite as many numbers as you need and eliminate the rest of the numbers, if the first number you drop is less than 5. In this case the first number you dropped is 2, so you rewrite 8.45 and drop 2,7. Let us now take the case of rounding 8.7694 to the hundredths place. When you round this the first number you drop is 9 which is more than 5. When this happens, you have to raise the preceding number by 1. Now 8.7694 drops the numbers 9, 4 and the 6 is raised to 7, and the rounded number becomes 8.77. Try your hand at a few rounding problems.

Problem. Round the following decimals to the hundredths place.
- (a) 6.898342
- (b) 3.24532
- (c) 5.34231

Answers
- (a) 6.90
- (b) 3.25
- (c) 5.34

Zero

The calculator will automatically display the results when you multiply, divide, add, or subtract with zero.

1. When a number is multiplied by zero, the answer is always zero.

Press	Display
30 × 0 =	0
0 × 30 =	0

2. When zero is divided by a number, the answer is zero.

Press	Display
0 ÷ 6 =	0

Dividing nothing (zero) into six parts is still nothing (zero).

3. Dividing by zero is not a legal operation mathematically.

Press	Display
8 ÷ 0 =	[Overflow is not legal.]

Notice this would overflow your calculator, since in math theory, any number divided by zero does not give an answer.

4. When you subtract zero from a number, the answer is the number itself.

---
*In this book we will round our decimals to the nearest hundredths place.

Press	Display
[15] [−] [0] [=]	15

5. When you add a number to zero, the answer will be the number itself.

Press	Display
[10] [+] [0] [=]	10

POWERS OF NUMBERS

The process of *raising a number to a power* is really a shortcut for multiplication of a series of like factors. For example, 3^4 or three to the fourth power is a shortcut way of expressing $3 \times 3 \times 3 \times 3$. The smaller number in the right hand corner is called an *exponent*. It tells you how many times you multiply the number. For example, 4^2 tells you to multiply 4 twice: 4×4.

Squaring

When you multiply a number by itself it is called *squaring*. You can square a number in your calculator by not entering it twice. You press [×] [=] and that squares any number.

Press	Display
[4] [×] [=]	16

Problem. Square the following numbers on your calculator: 9; 10; 12.

Answer

Press	Display
[9] [×] [=]	81
[10] [×] [=]	100
[12] [×] [=]	144

Now you know that any number can be squared (also referred to as *raising it to the second power* by pressing [×] [=]. If your calculator is in the constant mode and you continue to press the [=] key, this will continue to raise the number to the third, fourth, fifth, sixth and so forth powers. Each time you press the equal key it will increase the exponent or power by 1. Look at the following calculator examples.

Press	Display	Powers
[4] [×] [=]	16	4^2
[4] [×] [=] [=]	64	4^3
[4] [×] [=] [=] [=]	256	4^4
[4] [×] [=] [=] [=] [=]	1024	4^5

Problems

(a) 6^2
(b) 7^4
(c) 3^6
(d) 8^3

Answers

Problem	Press	Display
(a)	[6][×][=]	36
(b)	[7][×][=][=][=]	2401
(c)	[3][×][=][=][=][=][=]	729
(d)	[8][×][=][=]	512

Square Roots

When you square a number you multiply the number by itself. For example, $6 \times 6 = 36$. Finding the *square root* of a number is the opposite of squaring. You now find the quantity that, when multiplied by itself, gives the original number. For instance, the square root of 25 is 5 because 5 multiplied by itself equals 25. The calculator offers a fast way of finding square roots. You simply press the number to be squared followed by the square root key.

Problem. Find the square root of 25; 36; 5.

Answer

Press	Display
[25][√]	5
[36][√]	6
[5][√]	2.2360679

THE MEMORY KEY

Using the Memory Plus Key

Press [M+] whenever you want to put a number in *memory*. For example, to put 8 in memory (also referred to as *memory plus*) you push the following keys:

Press	Display
[8][M+]	8

When you put another number in memory, this will add the number to any number that you previously stored in memory rather than replacing the previous number. For instance, we have stored 8 in memory; if we now store 6 by pressing [6] [M+] the calculator will automatically add these two numbers together. Now if you recall memory by pushing the RM/CM key, you will recall the answer 14. When you work with a series of numbers and you want to add what is in memory you simply push the plus key and then the recall memory key. For example, if you have added 9 + 9 and you have on display 18 you now want to add what is in memory you would press the following keys:

Display	Press	Display
18	+ RM/CM	32

We will now put a few numbers in memory and then add memory to the value you will put on display.

Problem. Put the numbers 5, 6, 7, 8, 9, 20 in memory; then recall memory and add it to 50.

Answers

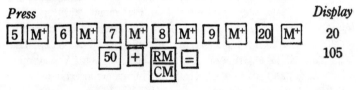

Press	Display
5 M⁺ 6 M⁺ 7 M⁺ 8 M⁺ 9 M⁺ 20 M⁺	20
50 + RM/CM =	105

Using the Memory Minus Key

When you press *memory minus* this puts a number in memory. For example using the same example if you put 8 in memory minus you push the following keys:

Press	Display
8 M⁻	8

When you put another number in memory minus, like memory plus this will add the number to any number that you previously stored in memory rather than replacing the previous number. Taking the same example, we have stored 8 in memory minus; if we now store 6 by pressing 6 M⁻, the calculator will automatically add these two numbers together. If you recall memory you will get the answer 14. Now you are probably asking why have memory minus; it seems to be the same as memory plus. But this is not true. The difference between memory plus and memory minus is when you have a number in display and add it to memory minus the value is *subtracted*. For example, when you have a display of 18 and when you add this to memory minus the answer is 4 instead of the 32 you got when you added memory plus. Press the keys and see.

Press	Display
[18] [+] [RM/CM] [=]	4

Now we will put a few numbers in memory minus and then add memory minus to the value you will put on display.

Problem. Put the numbers 6, 8, 9, 2 in memory minus, then recall memory minus and add it to 33.

Answers

Press	Display
[6] [M⁻] [8] [M⁻] [9] [M⁻] [2] [M⁻]	2
[33] [+] [RM/CM] [=]	8

THE NUMBER LINE

Positive and negative numbers can be best understood by looking at a number line. Let us look at the number line which follows:

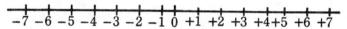

This line is marked off in equal spaces with a zero in the middle. Your *positive* numbers are on the right of the zero and your *negative* numbers are on the left. The sign of the number shows whether the number is to the right or left of the zero. In other words, a +7 tells you the 7 is to the right of the zero and a −7 tells you the number is to the left of the zero. The size of the number indicates the distance from zero. Addition and subtraction can be done on this line. For example, if you wish to add 2 + 4, you start at +2 and move 4 spaces to the right, ending on the answer: 6.

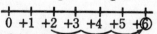

Subtraction is doing the reverse—moving to the left. If you want to subtract 2 from 7; start on 7 and move 2 spaces to the left, ending on 5.

It is very easy to perform these operations on a calculator and you don't have to use the number line. Let us say you wish to add 3 +5.

Press	Display
[3] [+] [5] [=]	8

There is no problem here; these are positive numbers. Now try to add 8 and −3. You enter the number 8 as usual: press the [+] function key and the number to be added (3)—but it *isn't* 3, so you change its sign by pressing [SC] and then you press the equal sign to get the answer 5. This example follows:

Press   *Display*
[8] [+] [3] [SC] [=]   5

Problems
1. −8 + −4 =
2. −4 + −2 =
3. 4 + −1 =

Answers

[8] [SC] [+] [4] [SC] [=]	−12
[4] [SC] [+] [2] [SC] [=]	−6
[4] [+] [1] [SC] [=]	3

Rules for Negative and Positive Numbers

There are certain rules for negative and positive numbers. These rules will be put here for easy reference. However, they really are not necessary when you are using your calculator. The calculator does all the thinking for you.

Adding Sign Numbers

1. If you have numbers of the same sign such as +8 + +4 or −5 + −6, you add the numbers with the same sign and put this sign on the sum. For example, +8 + +4 = +12 and −5 + −6 = −11. Use your calculator as follows, push each number in, change the sign when necessary, and push the function key. Both examples follow.

Press   *Display*
[8] [+] [4] [=]   12
[5] [SC] [+] [6] [SC] [=]   −11

Notice the calculator does all your thinking.

2. When you add numbers of different signs such as (−9) +2, the rule is you subtract the smaller number from the larger and put the sign of the *larger* on the answer. For example −9 + +3 = −6. The answer is −6 because the sign of the larger number 9 is minus. When doing this on a calculator, you again push each number in, change the sign when necessary, and push the plus function key. The example follows:

Press	Display
9 SC + 3 =	−6

3. When there is a whole series of negative and positive numbers* (for example $-8 + -3 + -2 + 1 + 5 + 6$) first add all the positive numbers ($1 + 5 + 6 = 12$), then add all the negative numbers ($-8 + -3 + -2 = -13$). We will now combine the results, $-13 + 12 = -1$. The answer is a -1 because the sign of the larger number is negative. (Refer to rule 2).

On the calculator this is much easier, as you need not do any of these things. You simply press the keys as is with the necessary change of sign when applicable. The calculator does everything. The above example is now shown on the calculator.

Press	Display
8 SC + 3 SC + 2 SC + 1 + 5 + 6 =	−1

Rules for Subtraction

4. When you subtract, you change the sign of the second number, then change the operation to addition and follow the appropriate rules for addition. For example, $21 - (-2)$ becomes $21 + 2 = 23$. We solve this problem using rule one. On the calculator none of this is necessary. When you solve this problem you enter the numbers and signs as they are and the calculator follows the rules. Let's show the sample example on the calculator.

Press	Display
21 − 2 SC =	23

Sign Numbers and Multiplication

5. Multiplication of numbers with the same sign always gives a positive product. For example, $6 \times 5 = 30$ and $-9 \times -4 = 36$. On the calculator you simply press the keys as is with the necessary change of sign when applicable. The calculator will do all the figuring.

Press	Display
6 × 5 =	30
9 SC × 4 SC =	36

6. Whenever you multiply unlike signs, the sign your answer will have is negative. For example, $-4 \times 6 = -24$. On the calculator you simply press the keys as is pushing a change of sign key whenever you have a negative number.

―――――――――
*A positive number may be written with or without a sign.

Press	Display
4 SC × 6 =	−24

Division of Sign Numbers

The rules you follow for division are the same as multiplication.

7. The same sign numbers will give you a positive answer: $\frac{6}{3} = 2$ and $\frac{-8}{-4} = 2$. Again, on the calculator you simply press the keys as is, pushing a sign key whenever you have a negative number. The examples on the calculator follows.

Press	Display
6 ÷ 3 =	2
8 SC ÷ 4 SC =	2

8. If you have a division problem with unlike signs (− 10 ÷ 5 = − 2), the answer will have a negative sign. Again on the calculator you simply press the keys as is pushing a change of sign key whenever you have a negative number. The example on the calculator follows.

Press	Display
10 SC ÷ 5 =	−2

ARITHMETIC OPERATION ORDER

There are definite rules of order when performing arithmetic operations. They are stated as follows:

1. If your numbers are inside a bracket, parentheses, or brace, for example (3+1) the operation should be performed first on these numbers.

2. If you have operations that involve exponents or square roots, for example 3^4 or $\sqrt{25}$, you perform operations on these numbers next.

3. After you have followed rules 1 and 2, then next do all the multiplications and divisions, then the additions and subtractions. Using your calculator, let us do a sample problem.

$$5 \times 4 + \sqrt{25} - 2 + 18 \div 2 + (8-2).$$

Working with the number in parentheses, we subtract 2 from 8 and get 6.

Press	Display
8 − 2 =	6

Our problem now looks like this: $5 \times 4 + \sqrt{25} - 2 + 18 \div 2 + 6$.

We now find the square root of $\sqrt{25}$ which is 5.

Press	Display
25 √	5

Our problem now looks like this: $5 \times 4 + 5 - 2 + 18 \div 2 + 6$. Next, we do all multiplications and divisions.

$$5 \times 4 = 20;\ 18 \div 2 = 9$$

Press	Display
5 × 4 =	20
18 ÷ 2 =	9

This leaves us with $20 + 5 - 2 + 9 + 6$. We add the numbers that are remaining in the problem.

Press	Display
20 + 5 + 9 + 6 =	40

Now you subtract 2.

Press	Display
40 − 2	38

The answer is 38. Now try your hand at a problem.

Problem. $8 \times 4 + \sqrt{100} - 3 + 18 \div 3 + (9 - 4) + 4^2$.

Answer

$$\begin{aligned}
8 \times 4 + \sqrt{100} - 3 + 18 \div 3 + (9-4) + 4^2 &= 8 \times 4 + \sqrt{100} - 3 + 18 \div 3 + 5 + 4^2 \\
&= 8 \times 4 + 10 - 3 + 18 \div 3 + 5 + 16 \\
&= 32 + 10 - 3 + 6 + 5 + 16 \\
&= 69 - 3 \\
&= 66
\end{aligned}$$

READING A STATISTICAL FORMULA

When you read a *statistical formula* you need to know what the symbols mean in the formula and the rule of arithmetic order. Some common symbols that you will encounter are as follows.

Symbol	Meaning
$\overline{X}$	Mean
Σ	Sum of
X	Raw score
$\sqrt{}$	Square root
N	Total number of scores
s	Standard deviation
X^2	Raw scores squared

In a statistical formula, the problem is always stated to the left of the equal sign and the directions are given to the right. For instance, in the raw score formula for the standard deviation

$$s = \frac{1}{N} \sqrt{N\Sigma X^2 - (\Sigma X)^2}$$

the problem is to find s (that is, the *standard deviation*) which is to the left of the equal sign and the directions are

$$\frac{1}{N} \sqrt{N\Sigma X^2 - (\Sigma X)^2}$$

which is to the right of the equal sign. The first thing we do is look at the directions and try to translate the symbols into commands. Pretend you have the scores 6, 2, 3, 4, 2, 1, 3, 2, 5, and 6, and you want to find the standard deviation. Looking at the directions for the standard deviation we see $\frac{1}{N}$; we know N represents the total number of scores. We count the number of scores for our example and you have 10. Therefore we can change this symbol of N to 10 wherever there is an N in the formula. The results follow.

$$s = \frac{1}{10} \sqrt{(10)\Sigma X^2 - (\Sigma X)^2}$$

The next symbol we see is ΣX^2 looking at the chart we see this means the sum of the squared raw scores. You square the individual scores: $6^2, 2^2, 3^2, 4^2, 2^2, 1^2, 3^2, 2^2, 5^2, 6^2$, then add the squared values: $36 + 4 + 9 + 16 + 4 + 1 + 9 + 4 + 25 + 36 = 144$; ΣX^2 is 144. The calculator handles this function much easier. You square each score and add the squared value to memory plus. You then recall memory plus to get the results. This procedure follows.

Press	Display
6 × = M⁺	36
2 × = M⁺	4
3 × = M⁺	9
4 × = M⁺	16
2 × = M⁺	4
1 × = M⁺	1
3 × = M⁺	9
2 × = M⁺	4
5 × = M⁺	25
6 × = M⁺	36
RM/CM	144

Wherever there is a ΣX^2, you are able to write 144. Your formula now looks like this:

$$s = \frac{1}{10} \sqrt{(10)144 - (\Sigma X)^2}$$

In the above formula, the only remaining symbol to be translated is $(\Sigma X)^2$ which is enclosed in parentheses. What this symbol tells you to do is to add the raw scores: $6 + 2 + 3 + 4 + 2 + 1 + 3 + 2 + 5 + 6 = 34$ and then square our sum: $34^2 = 1156$. The calculator procedure follows.

Press

6	+	2	+	3	+	4	+	2	+
1	+	3	+	2	+	5	+	6	=
×	=

Display

34
1156

The answer is 1156, so wherever there is a $(\Sigma X)^2$ you write 1156. Your formula now looks like this:

$$s = \frac{1}{10} \sqrt{(10)144 - 1156}$$

Now all that is necessary is for you to follow the arithmetic rules of order and solve this arithmetic problem. *1*. You multiply 10 by 144 ($10 \times 144 = 1440$). *2*. You subtract 1156 from 1440 ($1440 - 1156 = 284$). *3*. Square root this answer (16.852299). *4*. Divide the results (16.852299) by 10 ($16.852299 \div 10 = 1.6852299$). The answer is 1.69 rounded to the hundredths place. The calculator procedure is shown.

Press   *Display*
1. 144 × 10 =   1440
2. − 1156 =   284
3. √   16.852299
4. ÷ 10 =   1.6852299

Problem. Find the standard deviation of the following scores:

$$4, 5, 8, 10, 20, 3, 4, 2, 8, 10$$

using the formula $s = \frac{1}{N} \sqrt{(N)\Sigma X^2 - (\Sigma X)^2}$

Answer.

$$s = \frac{1}{N} \sqrt{(N)\Sigma X^2 - (\Sigma X)^2}$$

$$= \frac{1}{10} \sqrt{(10)798 - 5476}$$

$$= \frac{1}{10} \sqrt{2504}$$

$$= \frac{1}{10} \; 50.039984$$

$$= 5.00.$$

Now that you have studied this chapter, take Test I and see how well you do. The answers for the test follow it. If you miss a question, review the material.

TEST I

1. Solve the following decimal problems.

   (a) 89.23
        34.33
        90.23
      + 3.44

   (c) $\dfrac{.09}{.9} =$ _____

   (b) 7892.3
       − 194.2

   (d) 453.2
      ×   5.6

2. Round these numbers to the nearest hundredths place.

   (a) 19.2567 _____
   (b) 235.35689 _____
   (c) 82.34567 _____
   (d) 1.823 _____
   (e) 3.543 _____
   (f) 28.325 _____

3. Convert the following fractions to decimals.

   (a) $\dfrac{1}{5}$ _____

   (c) $\dfrac{1}{7}$ _____

   (c) $\dfrac{1}{6}$ _____

   (d) $\dfrac{1}{20}$ _____

4. Solve these problems:

   (a) $854.3 \times 0 =$ _____
   (b) $\dfrac{0}{9} =$ _____
   (c) $15 + 0 =$ _____
   (d) $543 \times 0 =$ _____

5. Solve the following problems.

   (a) $6^3 =$ _____
   (b) $\sqrt{225} =$ _____
   (c) $\sqrt{81} =$ _____
   (d) $9^7 =$ _____
   (e) $11^2 =$ _____
   (f) $\sqrt{49} =$ _____

6. Put the numbers 7, 3, 2, 1 in memory plus, then add memory plus to 65. What is your answer?

7. Put the numbers 15, 5, 3, 2, 1 in memory minus, then add memory minus to 70. What is your answer?

8. Complete the following operations:

   (a) $(-7) + (-10) =$ _____
   (b) $18 - (-7) =$ _____
   (c) $(-8) \times 5 =$ _____
   (d) $(-10) + 7 + 5 + (-3) =$ ___
   (e) $19 - 6 =$ _____
   (f) $\dfrac{-20}{10} =$ _____
   (g) $(-10) \times 8 =$ _____
   (h) $30 \times (-2) =$ _____

9. Solve the following problems.

   (a) $(9-8) + 5 - 3 \times 8 \div 4 =$ _____
   (b) $5^3 + 20 \div 5 + 8\sqrt{81} =$ _____
   (c) $5 \times 4 + 3 \times 6 \div 3 + 3 =$ _____

10. Use the formula $s = 1/N \sqrt{(N)\Sigma X^2 - (\Sigma X)^2}$ find the standard deviation for the following raw scores: 1, 9, 8, 5, 4, 2.
   $s =$ _____

ANSWERS

1. (a)   217.23
   (b)   7698.1
2. (a)   19.26
   (b)   235.36
   (c)   82.35
3. (a)   .20
   (b)   .1666666
4. (a)   0
   (b)   0
5. (a)   216
   (b)   15
   (c)   9
6.   78
7.   44
8. (a) -17
   (b)   25
   (c) -40
   (d) -1

   (c)   .1
   (d)   2537.92
   (d)   1.82
   (e)   3.54
   (f)   28.33
   (c)   .1428571
   (d)   .05
   (c)   15
   (d)   0
   (d)   4782969
   (e)   121
   (f)   7
   (e)   13
   (f) -2
   (g) -80
   (h)   60

9. (a) 0
   (b) 201
   (c) 29

10. $s = \dfrac{1}{6}\sqrt{(6)\,191 - 841}$

    $= \dfrac{1}{6}\sqrt{1146 - 841}$

    $= \dfrac{1}{6}\sqrt{305}$

    $= \dfrac{1}{6} \cdot 17.464249$

    $= 2.91.$

Chapter 2
Getting Acquainted
with the Calculator

All the statistical techniques in this book are designed to be done with the standard function calculator similar to the one pictured in Fig. 2-1. Instructions are given based on the way most calculators operate. Your particular calculator may operate slightly differently, so be sure you understand its operation so you can adapt the step-by-step procedure to it.

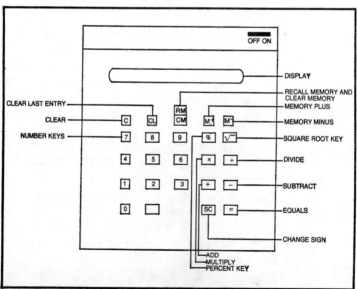

Fig. 2-1. Arrangement of typical calculator. Individual examples will, of course, vary.

Keys

0 – 9	Enters numbers. A maximum of eight digits (on most calculators) can be entered in the display.
.	Decimal point.
+	Instructs your calculator to add the next number you enter to the number in display.
−	Instructs your calculator to subtract the next number you enter to the number in display.
×	Instructs the calculator to multiply the displayed number by the next number you enter.
÷	Instructs the calculator to divide the displayed number by the next number you enter.
=	Instructs the calculator to finish the previously entered arithmetic operation and to put the results on display. Repeated use of = following a multiplication or division key is not disregarded by the calculator. The last number entry and multiplication or division key are reused by the calculator to perform a constant calculation. For example, 3 × = = = 81. In this case when the last number entry 3 and multiplication key are reused, the number 3 is multiplied by itself three times resulting in an answer of 81.
C	Clears the calculator and the display records 0. This key will not clear memory. Some calculators require the C key to be operated when the calculator is first switched on. Most calculators do not require you to clear them between calculations.
CL	Clears the last number entered on the keyboard. Sometimes referred to as CE.
%	Converts a number in display to the decimal equivalent by moving the decimal two places to the left. For example, 1200 × 6 % equals 72.
√	Finds the square root of the number in display.
M⁺	Adds a displayed number to the memory. Note that the M⁺ will add the number on display to any number that you previously stored in memory rather than replacing the previous number.
M⁻	Subtracts a display from memory. Note that M⁻ will subtract the number in display from any number that you previously stored in memory rather than

RM CM	replacing the previous number. A dual key. RM recalls the memory when the key is pushed once. When it is pushed a second time CM clears memory. Some calculators feature two separate keys with MR or RM recalling the memory and MC or CM clearing the memory.
SC	When change of sign is pressed after a number its sign changes. For example 3 SC becomes −3.

Now that you are familiar with your calculator we will learn how to compute a mean, standard deviation and standard error of the mean. The numbers used in the examples in this book are intentionally small to make the procedure easier to follow.

MEAN

A single score which describes a group of scores is called a *measure of central tendency*. The *mean* is one of the most popularly used measures of central tendency. To find a mean ($\overline{X}$) of raw scores, you sum the scores (ΣX) and divide by the number of scores (N). For example, you add the numbers 8, 4, 3, 2, 3, and divide by 5; the mean is 4. Notice for the mean *every* score affects the results. The mean is also considered the most stable of the measures of central tendency. That is, the mean of samples drawn from the population fluctuate only slightly. The mean is used to analyze data at the interval or ratio level.* If you compare the mean to other measures of central tendency such as the median or mode, it is the most affected by extreme scores when the distribution is small.

Example. A teacher wants to know the average score of the children participating in a special reading program. She records their scores. Find the mean for the following recorded scores: 6, 7, 8, 9, 10.

□ **Formula: Mean.**

$$\overline{X} = \frac{\Sigma X}{N}$$

Procedure	Record	Press	Display
1. Find N. Count the number of scores and			

---
*See Chapter 3 for a definition of interval and ratio.

Procedure	Record	Press	Display
write the answer in the column opposite $N = \_\_$.	$N = 5$		
2. Find ΣX. Add the scores. Press the following number and function keys.		$6 + 7 + 8 + 9 + 10 =$	40
3. Find the $\overline{X}$. Divide the sum of the scores (display) by the number of scores (record column)		$\div 5 =$	8
4. The mean equals 8.			

Practice Problems. What is the mean for the following scores?

(a) 10, 11, 9, 6, 2 _____
(b) 2, 4, 6, 8, 20, 12, 11 _____
(c) 50, 65, 33, 20, 26, 28, 16, 2 _____

Answers
(a) 7.6 (b) 9 (c) 30

STANDARD DEVIATION

A *measure of variability* is a single value which indicates the amount of variation in a group of scores. The *standard deviation* is a single value which indicates whether the numbers in a distribution are spread apart or clustered together around the mean. The standard deviation is the most widely used measure of variability. It is affected by every value in the distribution. It is considered the most stable of the measures of variability. That is, the standard deviation of samples drawn from the population fluctuate only slightly. It is used to analyze data at the interval or ratio level. If you compare the standard deviation to other measures of variability such as the range and quartile deviation, it is the most affected by extreme scores when the distribution is small.

Example. In a clinic for emotionally disturbed children one psychologist was recording the number of correct responses given by a sample of five children. Find the standard deviation for the following scores: 8, 9, 7, 6, 2.

☐ **Formula: Standard Deviation**

$$s = \frac{1}{N} \sqrt{(N)\Sigma X^2 - (\Sigma X)^2}$$

Procedure	Record	Press	Display
1. Find N. Count the number of scores and record.	N = 5		
2. Find ΣX^2. Square each score and add the squared value to memory. Next recall memory and see ΣX^2.		$\boxed{8} \times \boxed{8} \boxed{M^+}$	64
		$\boxed{9} \times \boxed{9} \boxed{M^+}$	81
		$\boxed{7} \times \boxed{7} \boxed{M^+}$	49
		$\boxed{6} \times \boxed{6} \boxed{M^+}$	36
		$\boxed{2} \times \boxed{2} \boxed{M^+}$	4
		$\boxed{RM}$ $\boxed{CM}$	234
3. Find $(\Sigma X)^2$. Add all the scores and square the sum. Write down the answer in the record column opposite $(\Sigma X)^2$ = ___.	$(\Sigma X)^2 = 1024$	$\boxed{8}+\boxed{9}+\boxed{7}+$ $\boxed{6}+\boxed{2}=\boxed{\times}=$	1024
4. Find $1/N \sqrt{(N)\Sigma X^2 - (\Sigma X)^2}$			
a. Multiply N (step 1) by ΣX^2 (memory).		$\boxed{5} \times \boxed{\begin{array}{c}RM\\CM\end{array}} \boxed{=}$	1170

Procedure	Record	Press	Display
b. Subtract $(\Sigma X)^2$ (step 3) from the display.		$\boxed{-}\boxed{1024}\boxed{=}$	146
c. Find the square root.		$\boxed{\sqrt{}}$	12.083045
d. Divide the square root by N (step 1) and the display is the standard deviation. $s = 2.416609$		$\boxed{\div}\boxed{5}\boxed{=}$	2.416609

Problems. What is the standard deviation for the following scores?

(a) 10, 11, 9, 6, 2 _____
(b) 2, 4, 6, 8, 20, 12, 11 _____
(c) 50, 65, 33, 22, 26, 28, 16 _____

Answers
(a) 3.2619012 (b) 5.5805784 (c) 15.970635

STANDARD ERROR OF THE MEAN

The *standard error of the mean* is thought of as a standard deviation of sample means. It is an estimate of how much the sample mean is expected to deviate from the population mean. The formula for the standard error of the mean is used only when the sampling distribution is normal or for a student t-distribution. Most of the data in the social sciences approximates the normal distribution. The normal distribution can be pictured as a bell shaped curve and looks as shown in Fig. 2-2.

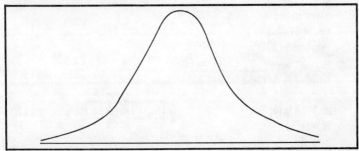

Fig. 2-2. Bell curve of normal distribution.

The t-distribution is more peaked than the normal distribution and it is not just one distribution. It looks like this:

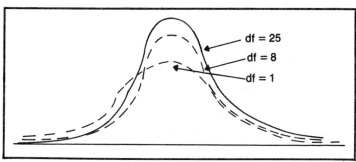

Fig. 2-3. Bell curve of t-distribution.

Example. Find the standard error of the mean for a sample of five emotionally disturbed children.

□ **Formula: Standard Error of the Mean**

$$S.E.M = \frac{s}{\sqrt{N}}$$

Instructions: The s used in the formula is the standard deviation. We will use the standard deviation 2.416609 and the N of 5 to find the standard error of the mean.

Procedure	*Record*	*Press*	*Display*
Find $\frac{s}{\sqrt{N}}$			
1. Put the standard deviation in memory.		2.416609 M⁺	2.416609
2. Find $\sqrt{N}$ Find the square root of N and record $\sqrt{N}$ =___.	$\sqrt{N}$ = 2.2360679	5 √	2.2360679
3. Find $\frac{s}{\sqrt{N}}$ Return from memory the standard deviation and divide this		RM/CM ÷ 2.2360679 =	1.0807404

25

Procedure	Record	Press	Display

by the $\sqrt{N}$ (step 2). The answer is the standard error of the mean.
S.E.M. = 1.0807404

Problems. Find the standard error of the mean for the following values:
(a) Given s = 2.34 N = 8
(b) Given s = 3.66 N = 10
(c) Given s = .66 N = 100

Answers
(a) .8273149 (b) 1.1573936 (c) .066

TEST 2

1. A single score which describes a group of scores is called a _____.

2. The mean is used to analyze data at the _____ level of measurement.

3. What is the standard deviation?

4. What is the standard error of the mean?

5. What does $\boxed{CL}$ instruct your calculator?

ANSWERS

1. Measures of central tendency.
2. Interval or ratio level.
3. The standard deviation is a single value which indicates the amount the numbers in a distribution are spread apart or clustered together around the mean.
4. It is an estimate of how much the sample mean is expected to deviate from the population mean.
5. Clear the last number entered on the keyboard. Sometimes referred to as CE.

Chapter 3
Review of Research Terms

This chapter will give you the background necessary to handle the statistical tests and techniques in the remaining chapters. The terms covered are as follows:

- [ ] Descriptive statistics
- [ ] Inferential statistics
- [ ] Population
- [ ] Sampling
    a. simple random
    b. cluster
    c. stratified
- [ ] Experimental study
- [ ] Independent variable and dependent variable
- [ ] Null hypothesis
- [ ] Alternative hypothesis
- [ ] Level of significance or alpha
- [ ] Sampling distribution
- [ ] Two-tail tests
- [ ] One-tail tests
- [ ] Degrees of freedom
- [ ] Parametric tests and nonparametric tests

DESCRIPTIVE STATISTICS

When the statistical methods you are using describe or give a clearer picture of the data they are called *descriptive statistics*. We call techniques such as the mean and standard deviation descriptive statistics because they are used for organizing and summarizing information that you observe.

27

INFERENTIAL STATISTICS

When you use statistical methods to infer beyond the data these methods are called *inferential statics*. These statistical methods attempt to draw conclusions about the population on the basis of observing only a smaller portion of this population. Statistical inference is primarily concerned with estimation of what a population is and testing hypotheses about a population. Some statistical techniques which are used to deal with these kinds of problems are the *Kruskal-Wallis* test, the *chi square* test, and the *one-way analysis of variance*.

POPULATION

A *population* is no more than a well-defined group of objects, people, plants or animals that all have something in common. For example, you may define a population as senior high school students at Clayton High in Clayton, Missouri in 1958. You are only then interested in senior high school students at Clayton High in Clayton, Missouri in 1958. You will *not* look at a population of senior high school students at Reseda High, in Reseda, California in 1958.

SAMPLING

It is often difficult or impossible to examine every member of a population. It is for this reason that we often use a smaller group called a *sample*. A sample is just a subgroup of the population. It can be small or large depending upon the study. An example of research that is dependent upon samples is the Neilsen ratings. These pollsters indicate the taste of the entire population of television watchers by using a selected sample. It is important that these samples are truly representative of the population. If the sample does not have the same elements as the original population, it is a biased sample and your study is not valid. We will now examine three methods of sampling.

Simple Random Sample

Every element of the population is listed when we take a *simple random* sample. From this list, a selection is made of a number of elements and this selection is our sample. For example, a business executive wants a sample of 300 men that work at his

---

*When we say *randomly selects* we mean that each member of the population will have the same opportunity of being selected.

plant. He first lists all 3,000 men in the population and from this list he would randomly* select 300 men for his sample. He must make sure every man has an equal chance of being selected. Also, the selection of one person must be independent of the selection of another person.

One of the best ways to insure a random sample is to use a device called a *table of random numbers*. There are other ways to insure randomness, such as a roulette wheel or a lottery.

Although simple random sampling is an important technique many studies do not use it. The reason is that to draw a simple random sample you must list ahead of time *every* item in the population. In many instances this is not feasible or practical and that is why we use other sampling techniques such as cluster sampling.

Cluster

Cluster sampling is just a variation of simple random sampling. However, this variation does not require that every element in the population be listed ahead of time. Cluster sampling is multistep random sampling. The population is divided into clusters and simple random samples of these clusters are then drawn. After this happens, the sample clusters are themselves divided into subclusters and a simple random sample is taken of the subclusters. This procedure continues until the researcher wants to stop. For example, if you are a pollster and you are using cluster sampling for the coming presidential election you might begin by looking at a map of fifty states or clusters. You use a table of random numbers to choose a sample of twenty states or clusters. Next, you divide these chosen states into congressional districts and draw a random sample of congressional districts. The next step is you divide those chosen congressional districts into voting precincts and draw a random sample of voting precincts. Finally you obtain a list of all the voters in these sample precincts and poll them.

Stratified Sampling

A *stratified* sample is again another variation of the simple random sample. You usually use a stratified sample when you want to make comparisons among different subgroups of the population or when you want to ensure that the sample is more representative of the total population. When you use a stratified sampling technique you must list all the elements of the population and have additional information on the relevant variable (s) in your population.* When you stratify a sample, you divide the population into homogeneous groups or strata based on relevant variables such as sex, social class or ethnic background. Each group has distinguishing characteristics that will set it apart from the other group. For example, a pollster might be interested in the voting preferences of people in Clay County. He wants to know who they will vote for in the coming election. He feels that certain characteristics will affect the voting for the only female candidate for mayor. These characteristics are sex, educational level, and age. Because of these possible influences, he feels that the proportion of these factors in the overall population must be represented in the sample.

The reason we have devoted so much time to a discussion of sampling is that the results based on a biased sample is worthless, no matter how beautifully your study is explained. (For further information on sampling consult Snedecor, listed in the suggested readings.)

EXPERIMENTAL STUDY

In an *experimental study* one investigates cause-and-effect relationships. You have two or more groups in the same situation except you vary the independent variable.

INDEPENDENT VARIABLE AND DEPENDENT VARIABLE

A *variable* is a characteristic that varies from one subject to another. A characteristic such as reading achievement is known as a variable. The independent variable is the one the researcher manipulates to see what changes occur in another variable. For example, if you are studying the effectiveness of a new teaching technique on childrens' achievement, the method of teaching is the independent variable. You manipulate this variable to see what effect it has on childrens' achievement. A dependent variable is the variable that fluctuates with the value of the independent variable, this variable responds to it. In the previous example, the independent variable is the childrens' achievement.

NULL HYPOTHESIS

We begin an experimental study by stating the *null hypothesis*. A hypothesis is a statement that offers a solution to a problem. The null hypothesis offers the solution there is no difference between

---
*A *variable* is a characteristic. This term will be explained in more depth shortly.

two distributions. For example, there is no difference between Group A and Group B. The null hypothesis is formulated with the intent of being rejected. When this occurs, the alternative hypothesis can now be accepted.

ALTERNATIVE HYPOTHESIS

The *alternative hypothesis* is the researcher's hypothesis. The researcher formulates the alternative hypothesis in two ways: (1) as a nondirectional hypothesis that says the two groups are unequal, or (2) as a directional hypothesis which predicts the direction of the difference between the two groups. In stating a directional hypothesis in order to reject the null hypothesis you must predict the direction of the difference ahead of time.

LEVEL OF SIGNIFICANCE OR ALPHA

The decision to reject or accept the null hypothesis is based on a certain probability which is called *alpha* or *level of significance*. Alpha is the probability of rejecting your null hypothesis wrongly. That is, your chances of being wrong. The level of significance that a researcher sets depends on his problem. In the social sciences the level of significance is usually .05 or .01.

SAMPLING DISTRIBUTION

A *sampling distribution* is a theoretical distribution that would occur if you randomly took all the possible samples of the same size from the population you were dealing with, computing a statistic for each sample. The statistics for these samples would form a distribution. Every statistical test has a sampling distribution. Some of the statistical tests use the same distributions.

TWO-TAIL TEST

A nondirectional test is often called a *two-tail test*.

ONE-TAIL TEST

Directional tests are referred to as *one-tail tests*.

DEGREES OF FREEDOM

Degrees of freedom refers to the number of values that are free to vary after restrictions have been placed on the data. The restrictions are inherent in the organization of the data. For example, if the data for 30 people are classified in two categories, and we know 20 people fall in one category, we know that 10 must fall in the other category. We enter many of the statistical tables in this book using degrees of freedom.

PARAMETRIC AND NONPARAMETRIC TESTS

Statistical tests can be labeled either *parametric* or *nonparametric*. In the book the parametric and nonparametric statistical tests are:

Parametric Tests	*Nonparametric Tests*
t(I)	Chi square (I)
t(II)	Chi square (II)
t(III)	McNemar
Randomized Groups	Cochran Q
Randomized Blocks	Kolmogorov-Smirnov
	Mann-Whitney U
	Wilcoxon
	Kruskal-Wallis
	Friedman

You should prefer using a parametric test because it is more powerful than a nonparametric test. That is, it will have a higher probability of rejecting the null hypothesis when it should be rejected. Four basic requirements are necessary before you can use a parametric test:

1. The groups are randomly drawn from the population.
2. The data has to be at least the interval level of measurement.
3. The data is normally distributed.
4. The variances are equal.*

Nonparametric tests have fewer and less stringent requirements. Requirement one for parametric tests is also applicable for nonparametric tests. The groups in the sample must be randomly drawn. The other assumptions are not met. The shape of the distribution does not have to be normal. The level of measurement is usually nominal or ordinal* (Chapter 4). When the sample size is extremely small you are definitely forced to use nonparametric tests.

TEST 3

1. A *variable* is _____
2. State a *null hypothesis*.

---

*Variance is the extent which scores cluster or scatter away from the mean in a distribution.

3. What are the two ways an *alternative hypothesis* can be formulated?

4. Why would you prefer using a *parametric* test over a *nonparametric* test?

5. A *two-tail* test is often called a _____ test.

6. _____ enables a person to go beyond his data from a sample.

7. _____ are used to summarize the data one is presently examining.

8. What is the difference between a *population* and a *sample*?

9. _____ occurs when the population is divided into homogeneous groups. A random sample is then drawn from each subgroup.

10. _____ is *multistep random sampling*.

ANSWERS

1. It is a characteristic that varies from one subject to another.

2. There is no significant difference between Group A and Group B.

3. a. as a directional hypothesis b. as a nondirectional hypothesis.

4. You would prefer using a parametric test because it is more powerful. That is, you have a higher probability of rejecting the null hypothesis when it should be rejected.

5. Nondirectional.

6. Inferential statistics.

7. Descriptive statistics.

8. A population is the entire group of objects or people, while a sample is only a part of the population.

9. Stratified sampling.

10. Cluster sampling.

---

*Refer to Chapter IV.

Chapter 4
How to Choose Quickly a Statistical Test

The accompanying statistical road map* (Fig. 4-1) gives the directions necessary for finding 14 of the commonly used statistical tests. These directions take into account the major requirements of each of these tests. Looking at this map you see boulevards, an intersection, avenues, streets and houses. The boulevards are called "Level of Measurement" and "Number of Groups." The major function of boulevards is to pose questions. The intersections are small rectangular shapes which are labeled "Nature of Groups." They also pose a question every time they are encountered.

The avenues which run between the boulevards are "Nominal," "Ordinal," and "Interval." The avenues major function is to answer the question posed by the "Level of Measurement" boulevard. The streets branch from the boulevards and intersection. Examples of streets are "Independent," "Related." The streets provide answers to questions posed by the boulevards and intersections. The houses on the map are where the statistical tests are located. Examples of some statistical tests are the Cochran Q and the Wilcoxon. Let's use the road map. Afterwards we will learn the reason for choosing one path over another, but for the moment let's operationally see how this map works. Starting at "Level of Measurement," this boulevard asks the question "What is the level of measurement?" You answer by traveling down "Nominal," "Ordinal," or "Interval" avenue. For the sake of this example, let us say your data is interval, so you choose "Interval"

---

*The map and explanation is adapted from *Statistics for the Social Sciences* by Vicki F. Sharp, Little, Brown and Company, 1979.

34

avenue. The next boulevard you encounter is "Number of Groups," which poses the question, "How many groups are there?" You can answer by choosing "1," "2" or "3 or More" streets. If you have two groups, you choose "2" street. You then reach the intersection "Nature of Groups." This intersection poses the question "What is the Nature of the groups?" You can answer this question by choosing "Independent" or "Related" street. If your groups are independent you choose "Independent" street and you will come to the statistical test called t(II).

In order to fully understand this map you must understand the meaning of certain terms. In the following paragraphs there is a short explanation of level of measurement, number of groups and nature of groups.

LEVEL OF MEASUREMENT

Level of measurement concerns the nature of the data. On the map there are three possible answers: *nominal, ordinal* or *interval*.

Nominal is only concerned with numbers that are assigned to identify or label persons or objects. A person's social security number is an example. The arithmetic operations you perform on these numbers is counting.

Ordinal concerns itself with numbers that are used to rank people or objects. The ranking of gemstones is an example of this. The arithmetic operations permissible are counting and ranking.

Interval allows you to compare persons or objects on a scale that has equal units. A standardized achievement test is an example of interval level of measurement. The arithmetic operations you perform on these numbers is addition and subtraction.

Ratio has an absolute zero; this permits multiplication and division. The Kelvin scale has a zero point that represents complete absence of temperature. Many of the studies in the social sciences do not require this measure, because they are not measuring a physical trait. Since there are few statistical tests that require this level, I have not included it on the map.

NUMBER OF GROUPS

You have three choices for number of groups. First, if you choose "1" you are looking at a single group. In the case of one group at the nominal level, it is still one group even though it may be subdivided into categories. For example, one group of second graders can be separated into two categories according to sex: male or female.

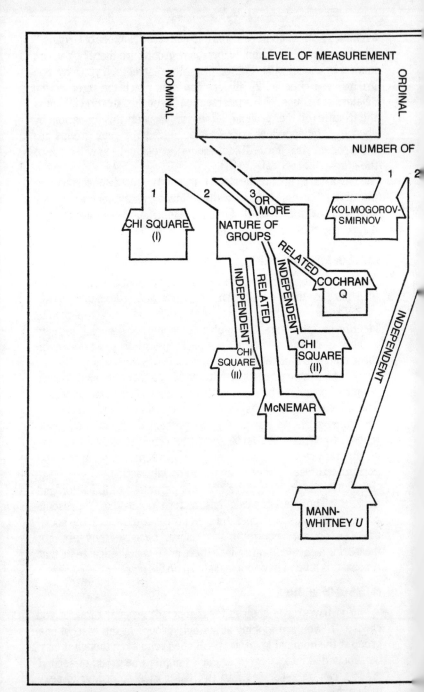

Fig. 4-1. A statistical road map.

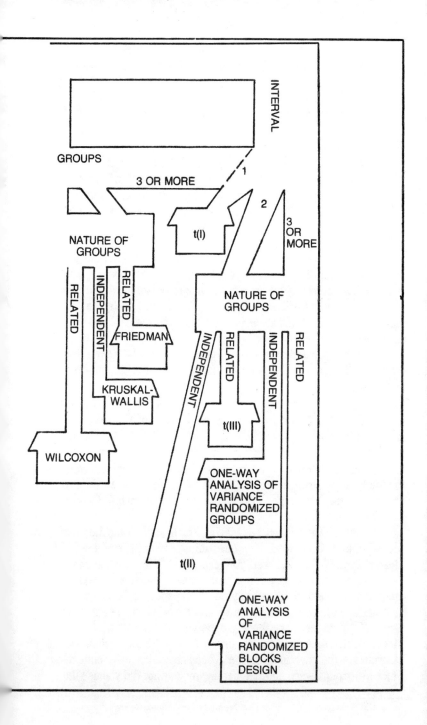

If you choose "2" you are interested in two groups. These two groups can be either two separate groups or the same groups used twice.

Last, if you choose "3 or more" you are interested in the differences between three or more groups.

NATURE OF GROUPS

What is the *nature of your groups*? Your answer can be "Independent" or "Related." Independent groups refer to groups that have members in their sample that are in no way connected to members in the other sample(s). The random samples* can be drawn from the same population or random samples from two different populations.

Related refers to group that have members that are matched with one another. The individual can be matched in two ways. You can match a person with himself. The same person can take two different treatments. The other way of matching is by pairing individuals in areas such as socio-economic status and IQ and then assigning them to each group.

Now that you understand the necessary terms, let's see if you can use the statistical map and solve the following practice problems.

The answer to the first practice problem will be discussed in some detail. This should provide you with a model of how to use the statistical road map.

Problem 1. A study was conducted to determine the preferences of 200 women for packaged candy. The subjects were randomly drawn from three income groups. They were asked to choose between wax paper, cellophane, and aluminum foil. Choose the correct statistical test.

Answer. The chi square test (II). You first go along "Level of Measurement" boulevard. The question is, "What is the level of measurement?" The answer you choose is "Nominal" avenue, because you are only recording preferences, which are frequencies; the women are deciding on wax paper, cellophane, or aluminum foil. The second boulevard traveled is "Number of Groups." Since there are three groups you choose to travel "3 or More," which takes you to "Nature of Groups" intersection. You ask yourself the question, are the groups related or independent? Since the groups were not matched in any way nor did you use the

---
*A random sample is a sample where each member of the population has the same chance of being selected.

same group twice; you then travel down "Independent" street to reach the statistical test chi square test (II).

Problem 2. You want to determine whether manufacturer A or manufacturer B produces better batteries. A random sample of 200 batteries are taken from Plant A and 200 batteries are taken from Plant B. Each group of batteries is tested and the exact time they last is recorded. Choose the correct statistical test.

Answer. t(II). You are recording your data in minutes and seconds which is ratio so you can easily choose "Interval" avenue. Since there are two groups, you choose "2" street. The nature of the groups are independent, since you have not matched the groups in any way nor did you use the same group twice. You then travel down "Independent" street to reach the statistical test t(II).

Problem 3. A group of 30 counseling students was formed into 15 matched pairs on the basis of a test which purports to measure empathy. One member from each matched pair was assigned at random to Group A; the other to Group B. Group A was given training in how to be an empathetic counselor while Group B was given instruction in some other unrelated topic. After six months, all the students were placed in a series of simulated social situations and were rated in empathy by judges on a scale from 1 to 5. Which test should be used to analyze this data?

Answer. The Wilcoxon test. The data is in the form of ranks, therefore the level of measurement is ordinal. There are two groups, you choose "2" street. The nature of the groups are related, since you have matched the groups on the variable empathy. You then travel down "Related" street to reach the Wilcoxon test.

The different sections of the Road map are covered by Chapters 5, 6, and 7. If your level of measurement is nominal, read Chapter 5 next. If your level of measurement is ordinal, turn to Chapter 6. If your level of measurement is interval, refer to Chapter 7. If you want a correlation technique consult Chapters 8 or 9.

TEST 4

1. A social security number is an example of _____ data.
2. A standardized achievement test is an example of _____ data.
3. Many studies in the social sciences do not require _____ level of measurement.
4. _____ groups refer to groups that have members in their samples who are in no way connected to members in the other sample(s).

5. Related groups are groups that have members that are matched to one another. How are these individuals matched?

ANSWERS
1. Nominal data
2. Interval data
3. Ratio
4. Independent
5. A person is matched with himself or individuals are matched in areas such as socioeconomic status and IQ.

Chapter 5
Nominal Tests

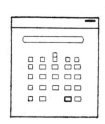

CHI SQUARE (I) TEST
Requirements.
1. Two-tail test only.
2. All the observations are used.
3. Nominal data.
4. It is only a one group test*.
5. One or more categories.
6. Sample size.
   a. Two categories the expected frequencies should be 5 or larger.
   b. More than two categories no more than 20% of the expected frequencies smaller than 5.
7. Simple random sample.
8. Data is in the form of frequencies.
□ **Formula: Chi Square.**

$$\chi^2 = \Sigma \; \frac{(O - E)^2}{E}$$

The chi square determines whether there is a significant difference between the expected and the observed frequencies. You determine the expected frequencies for chi square (I) in three ways:

1. If you hypothesize all frequencies are equal in each category, you figure the expected frequencies by dividing the number

---
*The author has taken certain liberties. The Chi square will be known as Chi square (I) when used with only one group and Chi square (II) when used with more than one group.

in the sample by the number of categories. For example, for a sample of 100 entering elementary school children, you would expect one half the entering children would be girls and one half would be boys. For this case the sample size is 100 and there are two categories, boys and girls. You divide the sample of 100 by 2 (the number of categories), and your answer is 50 expected frequencies for each category.

2. You determine expected frequencies on the basis of prior knowledge. Using our last example, pretend now we have prior knowledge of the number of boys and girls in each category for last year. At that time there were 65% boys entering and 35% girls. You find the expected frequencies by multiplying the new sample size by each hypothesized population proportion. If the new sample size is 100 you would expect 65 children to be boys (65% × 100) and 35 to be girls (35% × 100).

3. You base the expected frequencies on some theory. Mendel's famous theory is an example everyone likes to use. He discovered that when you crossed two dihybrid garden pea plants whose parents were hybrids, the ratio 9:3:3:1 results, that is, 9/16 of the pea plants were tall yellow, 3/16 were short yellow, 3/16 were tall green, and 1/16 were short green. If you crossed your own 200 pea plants using dihybrid garden peas like Mendel, your expected frequencies are determined by multiplying the sample size by each of the probabilities. In this instance, multiplying 200 (sample size) by 9/16 to obtain the expected frequency of 113 tall yellow pea plants*, multiplying 200 by 3/16 to obtain the expected frequency of 38 short yellow pea plants, multiplying 200 by 3/16 to obtain 38 tall green pea plants and multiplying 200 by 1/16 to obtain the expected frequency of 13 short green pea plants.

Example. Dick was a buyer for a supermarket chain. As a buyer he did not want to stock soap powder that was bought less frequently. According to Dick, the expected frequencies or number of customers choosing the soap powder should follow the percentages of last year. That would mean 30% should choose Brand A, 20% Brand B and 50% Brand C. He now took a random sample of 100 customers and asked their soap powder preference. The results of this poll are shown in the table below. Dick needs to know whether or not the discrepancies between last year's choices (expected frequencies) and this year's preferences on the basis of his poll (observed frequencies) demonstrates a real change in customer soap powder preferences.

---

*The numbers are rounded to the nearest whole number.

Soap Powder Category	O	E
Brand A	20	30
Brand B	40	20
Brand C	40	50

The *level of significance* is set at .05.

Null hypothesis states there is no significant difference between the expected and observed frequencies for the three brands of soap powder.

Alternative hypothesis states the expected and observed frequencies are different.

Procedure	Press	Display
1. Find $\frac{(O-E)^2}{E}$ for Brand A.		
a. You subtract the number in the E column from the number in the O column. $(O-E)$ and square the results. $(O-E)^2$	$\boxed{20}\ \boxed{-}\ \boxed{30}\ \boxed{\times}\ \boxed{=}$	100
b. Divide the number in display by its respective E value and put the results in memory plus to be added.	$\boxed{\div}\ \boxed{30}\ \boxed{M^+}$	3.3333333
2. Find $\frac{(O-E)^2}{E}$ for Brand B.		
a. You subtract the number in the E column from the number in	$\boxed{40}\ \boxed{-}\ \boxed{20}\ \boxed{\times}\ \boxed{=}$	400

Procedure	Press	Display
the 0 column. $(O - E)$ and square the results. $(O - E)^2$		
b. Divide the number in display by its respective E value and put the results in memory plus to be added.	÷ 20 M⁺	20
3. Find $\dfrac{(O - E)^2}{E}$ for Brand C.		
a. You subtract the number in the E column from the number in the 0 column. $(O - E)$ and square the results. $(O - E)^2$	40 − 50 × =	100
b. Divide the number in display by its respective E value and put the results in memory plus to be added.	÷ 50 M⁺	2
4. Find chi square by recalling memory. Write the answer in display in	RM CM	25.333333

44

| Procedure | Press | Display |

the record colum opposite χ^2 = ___. $X^2 = 25.333333$

Statistical Decision. Write your answers in the blanks to the left.

$\chi^2 = \underline{25.33}$ 1. Record your chi square value rounded to the hundredths place.

df = $\underline{2}$ 2. Find df. Subtract 1 from the number of categories. Write 2 opposite df = ___.

Table
Value = $\underline{5.99}$ 3. Find the table value. Use the df to enter Table A and run your fingers across this row until you are under your predetermined level of significance of .05. Write the value you find (5.99) opposite Table Value = ___.

reject   4. If your chi square value (25.33) is equal to or greater than the table value (5.99) reject the null hypothesis. Write reject.

There is a significant difference between the expected and observed frequencies for these three brands of soap powder. Therefore, Dick should pay attention to the results of this year's poll when he goes to stock his soap powder.

Problem 1. Carol did a survey of 160 of her customers on their frozen orange juice preference. The results of this poll are shown in the table below under the column labeled "observed frequencies." The expected frequencies are based on prior knowledge.

Level of significance: .01

Null hypothesis: there is no significant difference between the expected and observed frequencies.

Alternative hypothesis: they are different

Orange Juice Category	O	E
Brand A	30	40
Brand B	20	40
Brand C	40	40
Brand D	70	40

Answer. $X^2 = 35$. The table value is 11.34. Since the chi square value (35) is greater than the table value (11.34), reject the null hypothesis. The customers do not feel the same about all brands of orange juice.

Problem 2. Carlos did a survey of 100 of his customers on their hot dog preference. The results of this poll are shown in the following table under the column labeled "observed frequencies." The expected frequencies are based on the assumption the frequencies are equal in all the categories.

Level of significance: .05

Null hypothesis states there is no significant difference between the expected frequencies and the observed frequencies.

Alternative hypothesis states they are different.

Hot Dog Preference	O	E
A	40	25
B	10	25
C	20	25
D	20	25
E	10	25

Answer. $X^2 = 29$. The table value is 9.49. Since the chi square value (29) is greater than the table value (9.49), reject the null hypothesis. The customers do not feel the same about all brands of hot dogs.

Problem 3. Maria did a survey of 150 of her customers on their color choice of sweaters. The results of this survey are shown in the following table under the column labeled "observed frequencies." The expected frequencies are based on the assumption the frequencies are equal in all the categories.

Level of significance: .01

Null hypothesis states there is no significant difference between the expected frequencies and the observed frequencies.

Alternative hypothesis states they are different.

Color Preference	O	E
Red	28	30
Green	32	30
Blue	26	30
White	34	30
Yellow	30	30

Answer. $X^2 = 1.33$. The table value is 11.34. Since the chi square value (1.33) is *not* equal to or greater than the table value (11.34) we fail to reject the null hypothesis. The customers seem to feel the same about color preference.

CHI SQUARE(II) TEST

Requirements.
1. Two tail test only.
2. Nominal data.
3. Two or more groups.
4. Sample size should have no more than 20% of the expected frequencies smaller than 5.
5. Observations are independent.

Formula: Chi Square.

$$X^2 = \Sigma \frac{(O-E)^2}{E}$$

Example. The fifth grade principal of Clay Elementary School wanted to find out if there was a significant difference in the way his three fifth grade teachers gave pass/fail grades on a math exam to their respective classes. The data he recorded for all three teachers is shown below.

Teacher Grades

	Pass	Fail
John	25	7
Jack	10	6
Jill	11	6

The *level of significance* is set at .05.

The *null hypothesis* states there is no significant difference between the three classes.

The *alternative hypothesis* states the three teachers were not alike in the way they gave grades.

The expected frequencies for chi square (II) are derived from the data because, unlike chi square (I), there is *no* prior basis for computing them. We will first find the expected frequencies for our problem and then solve for chi square (II).

TABLE A
Teacher's Grades

	Pass	Fail	Row Total
John	25	7	32
Jack	10	6	16
Jill	11	6	17
Column totals	46	19	Grand Total 65

Instructions: We find the expected frequencies for the observed values in Table A and record the answers in Table B rounding to the nearest whole number.

TABLE B
Expected Frequencies

$\frac{46 \times 32}{65} = 23$	$\frac{19 \times 32}{65} = 9$
$\frac{46 \times 16}{65} = 11$	$\frac{19 \times 16}{65} = 5$
$\frac{46 \times 17}{65} = 12$	$\frac{19 \times 17}{65} = 5$

Procedure	Press	Display
1. Find the *row totals*. Add the numbers in each row and record the sum on the dashed line outside of Table A. Put each value in memory plus.	25 + 7 = M⁺ 10 + 6 = M⁺ 11 + 6 = M²	32 16 17
2. Find the *column totals*. Add the numbers in each column and place the sum on the dashed lines.	25 + 10 + 11 = 7 + 6 + 6 =	46 19
3. Find the *grand total*. Recall memory and put the display number in	RM CM	65

Procedure	Press	Display
the box labeled grand total.*		
4. Find the expected frequencies. Multiply each observed frequency's column total by its row total and then divide by the memory (grand total).	46 × 32 ÷ RM/CM =	22.646153
	46 × 16 ÷ RM/CM =	11.323076
	46 × 17 ÷ RM/CM =	12.030769
	19 × 32 ÷ RM/CM =	9.3538461
5. After you round, put the answers in Table B. For example, to find the expected frequency for John's observed of 25 you multiply his column total of 46 by his row total of 32 divide by 65 and round the answer 22.646153 to 23. Clear memory and display.	19 × 16 ÷ RM/CM =	4.676923
	19 × 17 ÷ RM/CM =	4.9692307
	RM/CM RM/CM C	0

Instructions: Record the observed frequencies from Table A and place them in the column labeled O. Next, record the expected frequencies from Table B and place them in the column labeled E. Now we will proceed as we did for the chi square (I).

---

*You can also find the grand total by adding column totals.

	Teacher's Pass/Fail Grades	O	E
John's Pass Grades		25	23
Jack's Pass Grades		10	11
Jill's Pass Grades		11	12
John's Fail Grades		7	9
Jack's Fail Grades		6	5
Jill's Fail Grades		6	5

Procedure	Press	Display

1. Find $\frac{(O-E)^2}{E}$

    John's pass grades.

    a. You subtract the number in the E column from the number in the O column. $(O-E)$ and square the results $(O-E)^2$.

$\boxed{25}\ \boxed{-}\ \boxed{23}\ \boxed{\times}\ \boxed{=}$	4

    b. Divide the number in display by its respective E value and put the results in memory plus to be added.

$\boxed{\div}\ \boxed{23}\ \boxed{M^+}$	.173913

You follow the same procedure for the remaining grades.

Procedure	Press	Display
2. Jack's Pass Grades	$\boxed{10}\ \boxed{-}\ \boxed{11}\ \boxed{\times}\ \boxed{=}$	1
	$\boxed{\div}\ \boxed{11}\ \boxed{M^+}$	.090909
3. Jill's Pass Grades	$\boxed{11}\ \boxed{-}\ \boxed{12}\ \boxed{\times}\ \boxed{=}$	1
	$\boxed{\div}\ \boxed{12}\ \boxed{M^+}$	.0833333

Procedure		Press	Display
4.	John's Fail Grades	[7][−][9][×][=]	4
		[÷][9][M+]	.4444444
5.	Jack's Fail Grades	[6][−][5][×][=]	1
		[÷][5][M+]	.2
6.	Jill's Fail Grades	[6][−][5][×][=]	1
		[÷][5][M+]	.2
7.	Find Chi square (II) Recall memory. Write the answer in display in the record column opposite $X^2 =$ ____.	[RM][CM]	1.1925997

$X^2 = 1.1925997$

Statistical Decision. Write your answers in the blank to the left.

$X^2 = \underline{1.19}$ 1. Record your chi square value rounded to the hundredths place.

$df = \underline{2}$ 2. Find df. Refer to Table A and use this formula (rows−1) × (columns −1). Write 2 opposite df = ____.

Table Value = $\underline{5.99}$ 3. Find the table value. Use the df to enter Table A run your fingers across the row until you are under your predetermined level of significance (.05). Write 5.99 opposite Table Value = ____.

<u>fail to reject</u>   4. If your chi square value (1.19) is equal to or greater than your table value (5.99) reject the null hypothesis.

The three teachers appear to be the same in the way they gave grades.

Problem 1. A sociology study compared three economic groups in their response to a question regarding party affiliation. The frequencies of response in each of three groups were compared.

Level of significance: .05.

Null hypothesis: there is no significant difference among the groups on their political party choice.

Alternative hypothesis: the groups are different

Economic Group	Democrat	Republican
High	30	10
Medium	8	22
Low	10	10

Answer. $X^2 = 16.89$.

Expected Frequencies

Economic Group	Democrat	Republican
High	21	19
Medium	16	14
Low	11	9

The *table value* is 5.99 (df = 2). The chi square value (16.80) is greater than the table value (5.99); therefore we reject the null hypothesis. The groups are not the same on political party choice.

Problem 2. A sample of children were classified according to their appearance and their adjustment to school. The children were then divided into three groups. The data are recorded in the following table:

Appearance	Adjustment to School		
	High	Medium	Low
Above Average	7	7	24
Average	16	8	8
Below Average	8	8	14

The *level of significance* is set at .01.
The *null hypothesis* states that the three groups are the same.
The *alternative hypothesis* states the groups are different.

Answer. $X^2 = 12.67$.

Appearance	Expected Frequencies Adjustment to School		
	High	Medium	Low
Above Average	12	9	17
Average	10	17	15
Below Average	9	17	14

The table value is 13.28 (df = 4). The chi square value (12.67) is not equal to or greater than the table value (13.28); we fail to reject the null hypothesis. The groups are the same.

Problem 3. A researcher investigating social class wanted to find out if high school students of different sociological backgrounds would enroll in different curriculums at Cal High School. She identified the social class membership and determined the curriculum enrollment for each student. The data follows.

	Social Class		
Curriculum	*I*	*II*	*III*
College	20	30	10
General	11	50	60
Vocational	6	25	40

The *level of significance* is set at .05.
The *null hypothesis* states that the three groups are the same.
The *alternative hypothesis* states the groups are different.
Answer. $\chi^2 = 32.98$.

	Expected Frequencies		
Curriculum	*I*	*II*	*III*
College	9	25	26
General	18	50	53
Vocational	10	30	31

The table value is 9.49 (df = 4). The chi square value (32.98) is greater than the table value (9.49); therefore, we reject the null hypothesis. The three groups are different.

McNEMAR TEST

Requirements

1. Two related groups.
2. Nominal data.
3. Expected frequencies 5 or more.
   You use the formula $E = \frac{1}{2}(A + D)$ to figure the expected frequencies.
4. Two-tail test only.

☐ **Formula: McNemar Test.**

$$\chi^2 = \frac{(|A - D| - 1)^2}{A + D}$$

Example. A survey of 30 voters was taken before a television advertising campaign. After the television film was shown, these same people were polled again. These individuals' responses are recorded in four possible ways, as shown in the table below.

After the Film

Before the Film	Democrat	Republican
Republican	16	4
Democrat	4	6

Instructions: A = number of individuals who change in one direction. D = number of individuals who change in the opposite direction from A. In this problem we will define A as equal to 16, the number of Republicans who change to Democrats. Therefore, D has to equal 6, the number of Democrats who change to Republicans.

The *level of significance* is set at .01.

The *null hypothesis* states that people's political attitudes do not significantly differ.

The *alternative hypothesis* states that people's attitudes are different.

Procedure	Record	Press	Display
1. Find the expected frequencies. $E = \frac{1}{2}(A + D)$			
a. Add A plus D and divide by 2.		$\boxed{16}\boxed{+}\boxed{6}\boxed{\div}\boxed{2}\boxed{=}$	11
b. If your answer is 5 or over proceed.			
2. Find the value for McNemar test.			
a. Subtract D from A.		$\boxed{16}\boxed{-}\boxed{6}\boxed{=}$	10
b. Subtract 1 from the display and square the value putting it in memory.		$\boxed{-}\boxed{1}\boxed{\times}\boxed{=}\boxed{M+}$	81

Procedure	Record	Press	Display
c. Add A plus D and put the answer in the record column opposite $A + D = $ ___.	$A + D = 22$	16 + 6 =	22
d. Recall memory and divide it by $A + D$ (step c).		RM/CM ÷ 22 =	3.6818181
e. Write the answer in the record column opposite $X^2 = $ ___.	$X^2 = 3.6818181$		

Statistical Decision. Write your answers in the blanks to the left.

$X^2 = \underline{3.68}$ 1. Record the value you found for the McNemar Test. Round it to the hundredths place.

Table Value = <u>6.64</u>   2. Find the table value. Enter Table A critical values of chi square with one degree of freedom.*
Run your fingers across the row until you are at your predetermined level of significance (.01). Write the value you find 6.64 opposite Table Value = ___.

<u>Fail to reject</u>   3. If your value in step 1 (3.68) is equal to or greater than the table value (6.64) reject the null hypothesis. Write fail to reject.

In this situation we fail to reject the null hypothesis. We can conclude that the voting preference of the people did not change significantly.

Problem 1. Maria took a random sample of 42 people and recorded whether they were for or against abortion. She then showed them a film favorable to abortion. One month later she recorded their position on this issue. Those people who changed from *for* to *against* she designated as A. Those who changed from

---
*Degrees of freedom is always 1 for the McNemar Test.

against to *for* she designated as D. The results are shown in the following table.

	After the Film	
Before the Film	Against	For
For	14	13
Against	10	5

The *level of significance* is .05.
The *null hypothesis* states that people's attitudes toward abortion remains the same.
The *alternative hypothesis* states that the people's attitude toward abortion is different.

Answer. $\chi^2 = 3.37$. The table value is 3.84. We fail to reject the null hypothesis. In this case, people's attitudes did not change.

Problem 2. Youngblood, a member of the school board, asked a group of 105 parents their feelings on busing. He then recorded their responses as for or against. One year later, after busing had taken place in the schools, he asked the same parents the same question and recorded their responses. The data follows.

	Before Busing	
After Busing	Against	For
For	20	40
Against	30	15

Those people who changed from *for* to *against* he designated as A.
Those who changed from *against* to *for* he designated as D.
The *level of significance* is .01.
The *null hypothesis* states that people's attitudes toward busing remains the same.
The *alternative hypothesis* states that the people's attitude toward busing is now different.

Answer. $\chi^2 = .46$ The table value is 6.64. We fail to reject the null hypothesis. In this case, people's attitudes did not change.

Problem 3. Pat Smith, a clinical psychologist, asked a group of psychologists their feelings on medication for slight depression. He recorded their responses as for or against. Six months later, after they took a course on the proper uses of drugs, he asked the same psychologists the same question and recorded their responses. The data follows.

| | Before Course | |
After Course	Against	For
For	40	10
Against	20	6

Those people who changed from *for* to *against* he designated as A. Those who changed from *against* to *for* he designated as D.

The *level of significance* is .05.

The *null hypothesis* states that the psychologists' attitudes toward medication remains the same.

The *alternative hypothesis* states that the psychologists' attitudes toward medication is now different.

Answer. $X^2 = 23.67$. The table value is 3.84. We reject the null hypothesis. The psychologists' attitudes toward medication is now different.

COCHRAN Q TEST

Requirements
1. Nominal data.
2. Three or more related groups.
3. Data in frequency form.
4. Two-tail test.

The scores for the Cochran Q test take only two values: 0 or 1. The 0 represents values that are recorded as negative, while the 1 represents values that are recorded as positive.

□ **Formula: Cochran Q Test.**

$$Q = \frac{(k-1)(k\Sigma C^2 - T^2)}{kT - \Sigma R^2}$$

Example. An efficiency expert devised three new methods to install engines in airplanes. He wanted to find out which method was easiest. He matched three groups of men on their ability to install engines. Group One was shown Method I, Group Two was shown Method II and Group Three was shown Method III. Each man was asked to write 0 if he felt the method shown was hard and 1 if he felt the method shown was easy. The data from this study is shown in the table that follows.

Level of significance is .05.

Null hypothesis states there is no significant difference among the three groups.

Alternative hypothesis states the groups are different.

	Method	Method	Method
Group	1	2	3
1	1	1	1
2	0	1	1
3	1	1	1
4	0	0	0
5	1	0	1

Instructions: You will put your answers in Box A, Box B or C, then in a summary table. Later you will transfer these answers to the Cochran Q formula.

Procedure	Record	Press	Display

$\boxed{A} = (k-1)$

1. Find k. Count the number of groups and write the answer opposite k = ___. $k = 3$

2. Subtract 1 from the number of groups and write the answer 2 in Box A. $\boxed{3 \mid - \mid 1 \mid =}$ 2

$$\boxed{A = 2} = (k-1)$$

Now transfer the answer (2) to the summary table under Box A.

Procedure	Record	Press	Display

$\boxed{B} = (k\Sigma C^2 - T^2)$

1. Find ΣC^2.

   a. Add the scores in the first column. Write your answer opposite $\Sigma C_1 =$ ___. $\Sigma C_1 = 3$ $\boxed{1 \mid + \mid 0 \mid + \mid 1 \mid + \mid 0 \mid + \mid 1 \mid =}$ 3

   b. Square your answer and put it in memory plus to be added. $\boxed{\times \mid = \mid M^+}$ 9

   c. Add the scores in the $\boxed{1 \mid + \mid 1 \mid + \mid 1 \mid + \mid 0 \mid + \mid 0 \mid =}$ 3

Procedure	Record	Press	Display
second column. Write your answer opposite $\Sigma C_2 = \underline{}$.	$\Sigma C_2 = 3$		
d. Square your answer and put it in memory plus to be added.		$\boxed{\times}\boxed{=}M^+$	9
e. Add the scores in the third column. Write your answer opposite $\Sigma C_3 = \underline{}$.	$\Sigma C_3 = 4$	$\boxed{1}\boxed{+}\boxed{1}\boxed{+}\boxed{1}\boxed{+}\boxed{0}\boxed{+}\boxed{1}\boxed{=}$	4
f. Square your answer and put it in memory plus.		$\boxed{\times}\boxed{=}M^+$	16
2. Find T^2.			
a. Find T. Add $\Sigma C_1 \Sigma C_2$ and ΣC_3 (refer to the record column opposite 1a, 1c and 1e). Write the answer in the record column opposite $T = \underline{}$.	$T = 10$	$\boxed{3}\boxed{+}\boxed{3}\boxed{+}\boxed{4}\boxed{=}$	10
b. Square the display and write it in the record column opposite $T^2 = \underline{}$.	$T^2 = 100$	$\boxed{\times}\boxed{=}$	100

Procedure	Record	Press	Display
3. Find $(k\Sigma C^2 - T^2)$. Multiply k (refer A_1 the record column) times the number in memory then subtract T^2 (refer to B_2 record column) and write your answer in Box B, then clear memory and display.		$\boxed{3}\boxed{\times}\boxed{\frac{RM}{CM}}\boxed{-}\boxed{100}\boxed{=}$	2
		$\boxed{\frac{RM}{CM}}\boxed{\frac{RM}{CM}}\boxed{C}$	0

$\boxed{B = 2} = (k\Sigma C^2 - T^2)$

Now transfer the answer (2) to the summary table under Box B.

Procedure	Record	Press	Display
$\boxed{C} = kT - \Sigma R^2$.			
1. Find ΣR^2. Add up the numbers in each row, square the answer and put it in memory.	$\boxed{1}\boxed{+}\boxed{1}\boxed{+}\boxed{1}\boxed{\times}\boxed{=}\boxed{M^+}$ $\boxed{0}\boxed{+}\boxed{1}\boxed{+}\boxed{1}\boxed{\times}\boxed{=}\boxed{M^+}$ $\boxed{1}\boxed{+}\boxed{1}\boxed{+}\boxed{1}\boxed{\times}\boxed{=}\boxed{M^+}$ $\boxed{0}\boxed{+}\boxed{0}\boxed{+}\boxed{0}\boxed{\times}\boxed{=}\boxed{M^+}$ $\boxed{1}\boxed{+}\boxed{0}\boxed{+}\boxed{1}\boxed{\times}\boxed{=}\boxed{M^+}$		9 4 9 0 4
2. Find $kT - \Sigma R^2$. *a*. Multiply k (refer to A_1) times T (refer to B_2) and subtract this from the value in memory. Write this answer		$\boxed{3}\boxed{\times}\boxed{10}\boxed{-}\boxed{\frac{RM}{CM}}\boxed{=}$	4

| Procedure | Record | Press | Display |

in Box C.
b. Clear memory and display. $\boxed{\frac{RM}{CM}} \boxed{\frac{RM}{CM}} \boxed{C}$ 0

$$\boxed{C = 4} = kT - \Sigma R^2$$

Now transfer the answer (4) to the summary table under Box C.

Instructions: We will now substitute in the formula

$$Q = \frac{(k-1)(k\Sigma C^2 - T^2)}{kT - \Sigma R^2}$$

the numbers from the Summary Table.

Summary Table

A	B	C
2	2	4

$$Q = \frac{(A)(B)}{C} \qquad Q = \frac{(2)(2)}{4}$$

| Procedure | Press | Display |

Find Q.
1. Multiply A times B and divide by C. $\boxed{2}\boxed{\times}\boxed{2}\boxed{\div}\boxed{4}\boxed{=}$ 1
2. The answer is the Cochran Q value. Q = 1.

Statistical Decision. Write your answers in the blanks to the left.

Q = <u>1</u>   1. Record the value you found for the Cochran Q Test. Round it to the hundredth place.

df = <u>2</u>   2. Find df. Use the formula (k − 1). Subtract 1 from the number of groups. Write 2 in the blank.

T = <u>5.99</u>   3. Find the table value. Enter Table A Critical Values of Chi Square with your df of 2. Run your fingers across this row until you are at your predetermined level of significance of (.05). Write the table value of 5.99.

<u>fail to reject</u>   4. If your value in Step 1 for Q is equal to or greater than the table value 5.99 reject the null hypothesis. Write fail to reject.

The three groups are not different in their feelings about the three methods of engine installation.

Problem 1. A cleaning expert devised four different methods of maintaining buildings. He wanted to find out which method was easiest for his organization to implement. He matched four groups of men on their ability to clean a building. Group One was then shown Method I, Group Two was shown Method II, Group Three was shown Method III, and Group Four was shown Method IV. Each person was asked to write 0 if he felt the method was hard and 1 if he felt the method was easy. The data is shown below.

Level of significance: .01.

Null hypothesis states that there is no significant difference among the four groups.

Alternative hypothesis states the groups are different.

Group	Method 1	Method 2	Method 3	Method 4
1	1	0	1	1
2	1	0	0	1
3	1	1	1	1
4	1	0	0	0

Answer. $Q = 6$. The table value is 11.34 (df=3). We fail to reject the null hypothesis in this case since the value for Q(6) is not equal to or greater than the table value (11.34). In this problem there was no significant difference in the methods that were tried.

Problem 2. A political pollster wanted to know if the dress of the interviewer influenced a housewife's response to an opinion survey on his political candidate. He matched three groups of housewives on the basis of their socioeconomic status. Group A was given a persuasive argument in favor of the candidate with the interviewer wearing very expensive clothes. Group B was given the same persuasive argument, but in this case the interviewer was wearing average clothes. Finally, Group C was given the same argument, but in this case the interviewer was wearing very shabby clothes. Each housewife was asked to write yes if she was in favor of the candidate and no if she was against him. The data was later recorded. A "no" was recorded as 0 and a "yes" as 1.

Housewife	Group A	Group B	Group C
A	0	0	0
B	1	0	1
C	0	1	1

Housewife	Group A	Group B	Group C
D	1	1	0
E	1	0	0
F	0	1	1
G	0	1	1
H	0	0	1
I	1	1	1

The *level of significance* is set at .05.

The *null hypothesis* states that there is no significant difference among the three groups.

The *alternative hypothesis* states the groups are different.

Answer. $Q = .86$. The table value is 5.99. (df = 2). We fail to reject the null hypothesis since the value for Q (.86) is not equal to or greater than the table value (5.99). In this case there was no significant difference. The clothing the interviewer wore did not appear to make a difference.

Problem 3. A manufacturer of signs shows three of his latest designs (A, B, and C) to eight buyers from neighboring retail stores. The buyers were asked to write yes if they were in favor of the sign and no if they did not like the sign.* A "no" was recorded as 0 and a "yes" as 1.

Sign Design

Buyer	A	B	C
1	0	1	1
2	0	1	1
3	1	1	1
4	0	0	1
5	0	0	1
6	0	0	1
7	0	0	1
8	0	0	1

The *level of significance* is set at .01.

The *null hypothesis* states that there is no significant difference in the demand for the three types of sign design.

The *alternative hypothesis* states the demand for sign design is different.

Answer. $Q = 11.14$. The table value is 9.21 (df = 2). The null hypothesis is rejected, because the Q value of 11.14 is greater than the table value of 9.21. The demand for sign designs is different.

---

*. This is a *related* test because each subject is exposed to every design.

TEST 5

1. Is the chi square test (I) a directional test?
2. A study was made to determine whether professors who were in the office on Friday when a researcher called differed in their opinion on integration from professors who were not there on Friday. To find out, a researcher interviewed those professors who were in on Friday and returned on other days to interview the rest. The data is shown below.

Professor Interviewed	Opinion Yes	No
Friday	125	10
Other Days	140	20

Which statistical test would you use and why?

3. With a value for $A = 5$ and a value for $D = 4$ can you use the McNemar test? Explain your answer.
4. What is the difference between the chi square test (I) and the chi square test (II)?
5. What distinguishes the Cochran Q test from the McNemar test?

ANSWERS

1. No.
2. You would use the chi square test (II), because the two groups are independent.
3. You cannot use the McNemar test. In order to use the McNemar test the expected frequencies must be at least 5. In this case the expected is 3.5 or 1/2 (5 + 4).
4. The chi square test (I) is a one group test, while the chi square test (II) is for two or more independent groups.
5. The Cochran Q test is an extension of the McNemar test. The McNemar test is used with only two groups, while the Cochran Q test is used with more than two groups.

Chapter 6
Ordinal Tests

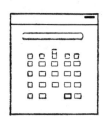

KOLMOGOROV-SMIRNOV TEST

Requirements
1. The data is ordinal.
2. One group test.
3. Simple random sample.

☐ **Formula: Kolmogorov-Smirnov Test**

$$D = \frac{LD}{N}$$

The Komogorov-Smirnov test compares the *cumulative distribution* of the observed scores and the cumulative distribution of the expected scores. Before you can use this test you must know how to create a cumulative distribution.

Forming a Cumulative Distribution. Once you are able to form a cumulative distribution, you will be able to form observed and expected cumulative distributions. The table below shows a cumulative distribution that has been created.

Rank	Frequency	Cumulative Distribution	Addition Work
1	2	20	6+8+4+2 = 20
2	4	18	6+8+4 = 18
3	8	14	6+8 = 14
4	6	6	6

You create a cumulative distribution by doing a series of additions on the frequency column. You simply add each number in the frequency column to the number(s) it has below it. You begin with 6 in the table; since there is no number below it, rewrite the 6 in the cumulative distribution column opposite the 6 in the frequency column. The next number is 8; since there is a 6 below it you add 6 + 8 and write 14 in the cumulative distribution column opposite 8 in the frequency column. You continue this procedure until you have formed the cumulative distribution column. If you have done this correctly the top number of your cumulative distribution column should equal the sum of the frequency column. That is, 20 should equal 2 + 4 + 8 + 6.

Example. Leonardo, a marketing specialist, wants to discern if the choice of tomatoes is affected by their color. To find out, he arranges to have several baskets of tomatoes placed before a group of twenty randomly selected customers. The tomatoes range from a very dark basket of tomatoes ranked as 1, to an extremely light basket of tomatoes ranked as 5. Each customer is then offered a choice among the five different baskets; the customers' selections are shown in the table below.

Rank	Observed Frequencies	Expected Frequencies
1	2	4
2	8	4
3	3	4
4	6	4
5	1	4

Instructions: The expected and observed frequencies are given. The expected frequencies are figured by dividing the number in the sample by the number of categories.

The *level of significance* is set at .05.

The *null hypothesis* states there is no significant difference in the preferences of the customers.

The *alternative hypothesis* states there is a difference in the customer's preferences. A two-tail test is used.*

Procedure	Press	Display
Find $D = \dfrac{LD}{N}$.		
1. Form the OC column. (Observed cumulative distribution.)	1 + 6 + 3 + 8 + 2 =	20
	1 + 6 + 3 + 8 =	18
	1 + 6 + 3 =	10
	1 + 6 =	7
	1 + 0 =	1

---

*The Kolmogorov-Smirnov test is a two-tail test except under special circumstances (Goodman L. A. "Kolmogorov-Smirnov, Tests for Psychological Research." Psychological Bulletin 51 (1954): 160-168.

Procedure	Press	Display
Write this column in Box A.		
2. Form the EC column. (Expected cumulative distribution.) Write this column in Box A.	4 + 4 + 4 + 4 + 4 = 4 + 4 + 4 + 4 = 4 + 4 + 4 = 4 + 4 = 4 + 0 =	20 16 12 8 4
3. In Box A form the OC - EC column. Subtract the numbers in the EC column from those in the OC column and record these values in the OC - EC column.	20 − 20 = 18 − 16 = 10 − 12 = 7 − 8 = 1 − 4 =	0 2 −2 −1 −3

	BOX A	
OC	EC	OC - EC
20	20	0
18	16	2
10	12	−2
7	8	−1
1	4	−3

Procedure	Record	Press	Display
4. Find LD. Look at the numbers in Box A in the OC-EC column. Take the largest number whether it		3 M+	3

Procedure	Record	Press	Display

is positive or negative, and put this number in memory.

5. Find N. Add the numbers in the observed frequencies column. Write the answer in the record column opposite N = ___ . 　N = 20

　　　　　　　　　　　$\boxed{2}\boxed{+}\boxed{8}\boxed{+}\boxed{3}\boxed{+}\boxed{6}\boxed{+}\boxed{1}\boxed{=}$　　20

6. Find $D = \dfrac{LD}{N}$

Divide the number in memory by N (refer to 5). D = .15

　　　　　　　　　　　$\boxed{\dfrac{RM}{CM}}\boxed{\div}\boxed{20}\boxed{=}$　　.15

Statistical Decision. Write your answers in the blanks to the left.

D = <u>.15</u>　1. Record your Kolmogorov-Smirnov value.

Table Value = <u>.294</u>　2. Find the table value. Enter Table C run your finger down the N column until you are at the value for N (refer to the record column 5), 20. Go across this row until you are at your predetermined level of significance, .05. The table value is .294 which is recorded opposite Table value = ___ .

<u>Fail to reject</u>　3. If the absolute value found for D(.15) is equal to or greater than the table value found in step 3 (.294) reject the null hypothesis.* Write fail to reject in the blank to the left. The color of tomatoes does not seem to affect customer's selections.

Problem 1. Luis, a movie photographer, wanted to see if choice of a starlet was determined by preferences among shades of skin color. To test how systematic skin-color preferences are, our photographer arranges to have a photograph taken of one starlet and developed in such a way that he obtains four copies. Each copy differs slightly in darkness from the other so that they can be

---

*The absolute value of a number is its numerical value, disregarding its positive or negative sign.

ranked from darkest to lightest skin color. A group of agents are then asked to choose among the four prints of these photographs. The results are recorded.

Level of significance: .01.

Null hypothesis: there is no significant difference in the preference patterns of agents in their choice of photographs.

Alternative hypothesis states that there is a difference in agents' choice of photograph by color. A two-tail test is used.

Agents' Photograph Choice

Rank	Observed Frequencies	Expected Frequencies
1	40	25
2	10	25
3	45	25
4	5	25

Answer. $D = .2$. The table value is .163. We reject the null hypothesis. The value for $D(.2)$ is greater than the table value (.163). We conclude that our agents show significant preferences among skin colors.

Problem 2. Mrs. Burns, an optician, wants to see if choice of reading glasses is determined by the size of the frames. To test this hypothesis, she arranges to have six different frames of the same color and shape placed before 18 randomly selected customers. The frames range in size from very small, ranked 1, to an extremely large one, ranked 6. The customers are then asked to choose their favorite frame; their selections are shown in the table below.

The *level of significance* is set at .05.

The *null hypothesis* states there is no significant difference in the preference of customers' for frames by size.

The *alternative hypothesis* states there is a difference in customers' choice of frames by size. A two-tail test is used.

Customers' Choice of Frames

Rank	Observed Frequencies	Expected Frequencies
1	1	3
2	1	3
3	3	3
4	3	3
5	10	3
6	0	3

Answer. D = .22. The table value is .309. We fail to reject the null hypothesis. The value for D (.22) is not equal to or greater than the table value. The size of glass frames does not seem to affect customers' preference.

Problem 3. A special education teacher feels that emotionally disturbed children will feel comfortable when other children maintain a degree of physical distance. To test this, she asks a group of 25 disturbed children to look at five photographs and select the one they prefer. The photographs portray two children sitting at various distances from one another, otherwise the pictures are the same. The closest distance is given a rank of 1, and the farthest distance is given a rank of 5. The childrens' selections are recorded in the table below.

The *level of significance* is set at .01.

The *null hypothesis* states there is no significant difference in the preference patterns of children in their choice of photographs which portray different physical distance.

The *alternative hypothesis* states that there is a difference in childrens' choice of photographs which portray different physical distance. A two-tail test is used.

Childrens' Choice of Photograph

Rank	Observed Frequencies	Expected Frequencies
1	0	5
2	1	5
3	1	5
4	2	5
5	21	5

Answer. D = .64. The table value is .32. We reject the null hypothesis. The value for D (.64) is greater than the table value of .32. The distance of children from each other in the pictures does affect the childrens' photograph preferences.

MANN-WHITNEY U TEST

Requirements

1. The data is ordinal.
2. There are two independent groups.
3. Data is in ranks.
4. Simple random samples.
5. Sample size can be different for the two groups.

☐ **Formula: Mann-Whitney U**

$$U_1 = n_1 n_2 + \frac{n_1(n_1 + 1)}{2} - \Sigma R_1$$

$$U_2 = n_1 n_2 - U_1$$

Before you can use the Mann-Whitney U, you must know how to rank data. In fact many of the tests we will be doing rank their scores, so you should be familiar with this procedure.

How To Rank. In this volume, we will rank from lowest to highest. Suppose we have the following six scores: 3, 9, 6, 2, 5, 10.

1. We place these numbers in order from lowest to highest: 2, 3, 5, 6, 9, 10.

2. Next, we assign each value a rank number. The smallest (2) is assigned a 1; next (3) is assigned a 2; 5 is assigned a 3, and so forth until the largest (10) is assigned a six. The results of this ranking are shown:

Score	2	3	5	6	9	10
Rank	1	2	3	4	5	6

The only problem you encounter when you rank, occurs when numbers are tied. For example, if you are asked to rank 6, 5, 6, 6, 8, you will notice three numbers are tied for the same rank—there are three 6s. You handle the tie by assigning each tied value the average rank of the rank positions they occupy. Let me demonstrate. First, order the numbers 5, 6, 6, 6, 8. Now let us assign ranks. The number 5 is given a rank of 1. Since three 6s occupy rank positions 2, 3 and 4, we find the *average* rank by adding these three rank positions together and dividing by the number of rank positions involved in this case three. That is, $(2 + 3 + 4)/3 = 3$. You now give the 6s each the rank of 3. You assign the next number the rank of 5. The results follow:

Score	5	6	6	6	8
Rank	1	3	3	3	5

To make sure you understand ranking solve the practice problem.

Problem. Rank the following numbers: 22, 44, 21, 66, 2, 22, 30, 22.

Answer

Score	2	21	22	22	22	30	44	66
Rank	1	2	4	4	4	6	7	8

Example. Tony wanted to find out if there was a difference in the way groups of dancing students rated an experienced dancer.

He randomly selected two groups of prospective dancers, the experimental group (4 students) took a special dance training course, the control group (5 students) took the regular dance course. At the end of training, each group was shown a video tape of an experienced dancer and was asked to give him a score from 1 to 20 depending on his fulfilling certain requirements. The ratings are shown in the table below.

Experimental Group		Control Group	
n_1	R_1	n_2	R_2
10	2	16	5
12	3	18	7
2	1	20	9
17	6	19	8
		13	4

Instructions: Label the individuals in the smaller groups as n_1. When you rank the scores, combine the scores of both groups together.* You will put your answers in Box A, B or C for the value of U_1, then the summary table. Later, you will transfer these answers to the U_1 formula.

The *level of significance* is set at .05

The *null hypothesis* states there is no significant difference between the ratings of the two groups.

The *alternative hypothesis* states the two groups are different in the scores they give the experienced dancer. We will use a two-tail test.

Procedure	Record	Press	Display
1. $\boxed{A} = n_1 n_2$ Find n_1. Count the number of individuals in the group labeled n_1. Write the number opposite $n_1 = $ ___ in the record column.	$n_1 = 4$		

---

*For this problem, the scores have been already ranked and placed in the R_1 and R_2 columns. A check to see if you ranked properly is $\Sigma R_1 + \Sigma R_2 = N[(N+1)/2]$. N is defined as the number of subjects.

[1] The experimental group is given the treatment while the control group is not given any treatment.
[2] Use the smaller group regardless of whether it is the experimental or control group.

Procedure	Record	Press	Display
2. Find n_2. Count the number of individuals in the group labeled n_2. Write the number opposite $n_2 = \underline{}$ in the record column.	$n_2 = 5$		
3. Find $n_1 n_2$. Multiply n_1 times n_2 (refer to A_1 and A_2) and write the answer in display in Box A.		$\boxed{4}\ \boxed{\times}\ \boxed{5}\ \boxed{=}$	20

$$\boxed{A = 20}\ =\ n_1 n_2$$

Now transfer the answer (20) to the summary table under Box A.

Procedure	Record	Press	Display
$\boxed{B} = \dfrac{n_1(n_1 + 1)}{2}$			
1. Find $(n_1 + 1)$. Add 1 to n_1 (refer to A_1) then multiply it by n_1 and divide by 2. Write the answer in Box B.		$\boxed{4}\ \boxed{+}\ \boxed{1}\ \boxed{\times}\ \boxed{4}\ \boxed{\div}\ \boxed{2}\ \boxed{=}$	10

$$\boxed{B = 10} = \frac{n_1(n_1+1)}{2}$$

Now transfer the answer (10) to the summary table under Box B.

Procedure	Record	Press	Display
$\boxed{C} = \Sigma R_1$			
1. Add all the numbers in the R_1 column, and put the answer in memory plus; also write it in Box C.		$\boxed{2}+\boxed{3}+\boxed{1}+\boxed{6}\boxed{M^+}$	12

$$\boxed{C = 12} = \Sigma R_1$$

Now transfer the answer (12) to the summary table under Box C.

Instructions: We will now substitute in the formula

$$U_1 = n_1 n_2 + \frac{n_1(n_1+1)}{2} - \Sigma R_1$$

the numbers from the Summary Table.

Summary Table

A	B	C
20	10	12

$U_1 = A + B - C$
$U_1 = 20 + 10 - 12$

Procedure	Record	Press	Display
Find U_1.			
1. Add Box A and Box B; subtract Box C which is in memory. Write the answer opposite		$\boxed{20}+\boxed{10}-\boxed{\frac{RM}{CM}}\boxed{=}$	18

Procedure	Record	Press	Display
$U_1 = $ ___ in the record column and circle it.	$(U_1 = 18)$		

Procedure	Record	Press	Display
Find $U_2 = n_1 n_2 - U_1$			
1. Subtract the circled value (U_1) from Box A.		$\boxed{20}\ \boxed{-}\ \boxed{18}\ \boxed{=}$	2
2. Write the answer in display in the record column opposite $U_2 = $ ___.	$U_2 = 2$.		

Use Table C when n_2 is 8 or less. When n_2 is between 9 and 20, use Table D.

Statistical Decision. Write your answers in the blank to the left.

$U = \underline{2}$ 1. Compare U_1 and U_2 the smaller value is U write this opposite $U = $ ___ in the blank to the left.

$\alpha = \underline{.05}$ 2. Record your predetermined level of significance in the blank to the left. Write .05.

$T = \underline{.064}$ 3. Find the table value for U. Refer to Table C, since n_2 is 8 or less. Since your n_2 is equal to 5 refer to the table labeled $n_2 = 5$. Find your U of 2 in the left-hand column in this table. Run your fingers across the row until you are under your n_1 of 4. Where these points intersect is the exact probability of .032. These probabilities are one tail probabilities; since you need two tail probabilities, simply double the value found in the table and write this as your table value: $.032 \times 2 = .064$.

4. Reject the null hypothesis if your table value (step 3) is less than your predetermined level of significance (step 2). Since .064 is not less than .05 write fail to reject. There is no significant difference in the ratings of the two groups.

Problem 1. A dog owner is experimenting with rewards in training his dogs. He wants to see how effective these techniques are in getting dogs to obey vocal commands. He randomly selects 18 dogs. One group of 8 dogs is trained with no rewards, while the other group of 10 dogs is rewarded. The results are recorded in the table below.

Level of significance: .05.

Null hypothesis states there is no significant difference between the two groups of dogs.

Alternative hypothesis there is a difference between the two groups of dogs. A two-tail test is used.

n_1 Scores I	R_1	n_2 Scores II	R_2
20	14	15	11
13	9	18	12
2	2	24	15
11	8	19	13
4	4.5	27	16
6	6	33	18
14	10	30	17
1	1	4	4.5
		3	3
		8	7

Answer. $U = 18.5$. Table Value = 17. We fail to reject the null hypothesis. In this situation we use Table E.* There are four tables given. Since your problem is a two-tail test at the .05 level of significance, you have to choose the appropriate table. You run your fingers down the extreme left column of this table until you

---
*n_2 is between 9 and 20.

find n_1 (8). Next, run your fingers across this row until you are under n_2 (10). Record the table value of 17. If the value found for U(18.5) is *equal to* or *smaller than* the table value of 17, reject the null hypothesis. Since 18.5 is not smaller than 17, you cannot reject the null hypothesis. There is no significant difference between the two groups of dogs.

Problem 2. A movie producer was casting a new movie. He wanted to determine whether Lee singers or the Ruth singers were better to use for his picture. He felt the Ruth group is a better singing group, but they charged more money. The producer had a panel of experts rate each group on their singing ability. They rated them 1 (poor) to 25 (sensational). The results were recorded in a table.

Level of significance: .01.

Null hypothesis states there is no significant difference between the two groups of singers.

Alternative hypothesis states that the Ruth group is better than the Lee group in their singing ability. A one-tail test is used.

Singing Group Ratings

Lee		Ruth	
n_2	R_2	n_1	R_1
10	7	9	5
15	9	8	4
20	11.5	22	14
10	7	20	11.5
25	17	10	7
16	10	5	2
24	16	5	2
23	15		
21	13		

Answer. U = 11.5. The table value is 11. You cannot reject the null hypothesis. In this situation we again use Table E. Since the value found for U(11.5) is not equal to or smaller than the table value 11, you cannot reject the null hypothesis. The Ruth group is not significantly better than the Lee group.

Problem 3. A political scientist wants to find out if there is a significant difference in how people rate a political candidate after they have seen a film on propaganda. He randomly selects two groups of subjects. One experimental group of 7 subjects sees the propaganda film. The control group of 8 subjects just talked about politics. The subjects then had a politician come to talk to them. They were asked to rate him. The total ratings possible was 20. The table below indicates the results.

Level of significance: .05.

Null hypothesis states that the two groups are the same in their ratings of this politician.

Alternative hypothesis states the two groups are different in the scores they give the politician. A two-tail test is used.

Politician Effectiveness Ratings

Experimental Group		Control Group	
n_1	R_1	n_2	R_2
9	6	15	10
12	8	17	12
3	2	19	14
7	5	18	13
5	3	20	15
11	7	16	11
2	1	14	9
		6	4

Answer. $U = 4$. The table value is .004. You double the value in Table C since this is a two-tail test. You reject the null hypothesis, since the table value (.004) is less than the level of significance (.05). The two groups are different in the scores they gave the politician.

When n_2 is larger than 20. When n_2 is larger than 20 after you find U, you have to use another formula which is

$$z = \frac{U + \frac{1}{2} - \frac{n_1 n_2}{2}}{\sqrt{\frac{n_1 n_2 (n_1 + n_2 + 1)}{12}}}$$

To find the table value, you go to Table D after you find z. For example, let's say you have a z value of 2.5, a level of significance of .05, and a two tail test. You run your fingers down the z column till you come to 2.5, you then go across this row and under .00 you find a probability of .0062. For a two-tail test you must double this probability. Therefore .0062 becomes .0124. If this probability .0124 is *equal to or less* than your level of significance .05 you reject the null hypothesis. In this case you reject.

WILCOXON SIGNED-RANKS TEST

Requirements
1. The data is ordinal.
2. Two related groups.
3. Ranked data.

☐ **Formula: Wilcoxon Signed-Ranks Test**

$$T = \text{smaller value for either } \Sigma R^+ \text{ or } \Sigma R^-$$

Example. Mr. Bill Murphy, a head coach, has to choose between two football coaches who have opposite approaches. He matches 14 football players on their ability to execute certain plays. One player from each matched pair is assigned at random to one of the two coaches. For 80 hours each coach teaches the players new plays. At the end the players are tested and rated from 1 to 30 on their ability to execute the new plays. The results are expressed in the table below.

Pair	Column 1 Coach A	Column 2 Coach B
A	20	30
B	20	6
C	29	24
D	18	23
E	26	22
F	26	26
G	14	5

Level of significance is .05.

Null hypothesis states there is no significant difference between Coach A's group and Coach B's group.

Alternative hypothesis states there is a significant difference between the groups. A two-tail test is used.

Procedure	Record	Press	Display
Find T.			
1. Form column D. Subtract each number in column two from its corresponding number in column one. Next		$\boxed{20} - \boxed{30} =$ $\boxed{20} - \boxed{6} =$ $\boxed{29} - \boxed{24} =$ $\boxed{18} - \boxed{23} =$ $\boxed{26} - \boxed{22} =$ $\boxed{26} - \boxed{26} =$ $\boxed{14} - \boxed{5} =$	−10 14 5 −5 4 0 9

Procedure	Record	Press	Display

write these values in a column labeled D in Box A.
2. Find N. Count the number of D's that are not 0 and record this opposite N = ___.   N = 6

```
                    BOX A
        D       R       R⁺          R⁻
      -10      -5                   -5
       14       6       6
        5       2.5     2.5
       -5      -2.5                 -2.5
        4       1       1
        0
        9       4       4
                      ─────       ───────
                      Σ⁺=13.5     ΣR⁻=7.5
```

Procedure	Record	Press	Display

Form the R column.
3. Rank all the values in the D column that are not zeroes. When you rank disregard their signs. After you rank record the ranks in Box A. Next,

Procedure	Record	Press	Display
place the correct sign in front of the number in this column.			

Find the ΣR^+.
4. Add all the numbers that have a positive rank. Write the answer opposite $\Sigma R^+ =$ ___ in the record column.

$\Sigma R^+ = 13.5$

| 6 | + | 2.5 | + | 1 | + | 4 | = |

13.5

Find ΣR^-.
5. Add all the numbers that have a negative rank. Write the answer opposite $\Sigma R^- =$ ___ in the record column.

$\Sigma R^- = 7.5$

| 5 | SC | + | 2.5 | SC | = |

−7.5

Find T.
Compare the answers found in Step 4 and Step 5. Record the smaller number abso-

| Procedure | Record | Press | Display |

lute value)
disregarding
the sign T = 7.5
as T. T = 7.5

Statistical Decision. Write your answers in the blanks to the left.

T = <u>7.5</u> 1. Record the value you found for T opposite T = ___. Write 7.5.

N = <u>6</u> 2. Write the value you found for N (Step 2 record column) opposite N = ___. Record 6.

Table Value = <u>1</u> 3. Find the table value. Refer to Table F. Run your fingers down the N column until you are at the proper value for N(6). Next, go across the row until you are under the predetermined level of significance (0.5) and the proper tail (two-tail). Write the number you find as your table value opposite T = ___. Write 1.

<u>Fail to reject</u> 4. If the T value (Step 1) is *equal to or less* than the table value (Step 3) reject the null hypothesis. Since 7.5 is not less than 1 you fail to reject the null hypothesis. There is no significant difference between the two groups.

Problem 1. Maria wants to find out if music has an effect on plants. She buys two sets of identical plants and assigns one plant from each matched pair to a room. The rooms are identical. For three months the plants are given the same treatment except Group A has music played two hours a day. At the end of the experiment Maria has an expert called in to rate the plants on a scale from 1 to 50. The results are shown in the accompanying table.

Level of significance is .05.
Null hypothesis there is no significant difference between Group A and Group B.
Alternative hypothesis states that Group A will do better than Group B. A one-tail test is used.

Pair	Music Group A	Nonmusic Group B
A	39	30
B	50	20
C	23	33
D	18	11
E	11	9
F	4	3

Answer. T = 5. The table value is 2. Remember this is a one-tail test. We fail to reject the null hypothesis, since the T value (5) is not *equal* to or *less* than the table value (2). The plants for both groups are the same.

Problem 2. A human relations expert wants to test the effect of a human relations course on a group of students. For his study he is fortunate enough to have eight pairs of identical twins. At random, one twin from each pair is assigned to attend a basic course in psychology and the other to attend a psychology course specializing in human relations techniques. At the end of the term, the sixteen students are rated to determine their ability to be sociable. The results are expressed in the following table.

Pair	Basic Course Group	Human Relations Group
A	15	20
B	25	7
C	23	27
D	18	11
E	11	4
F	29	10
G	10	10
H	29	18

Level of significance is .01.

Null hypothesis: there is no significant difference between the human relations groups and the basic course group.

Alternative hypothesis states there is a difference between the two groups. This is a two-tail test.

Answer. T = 3. The table value is 0. We fail to reject the null hypothesis, since the T value (3) is not equal to or less than the table value (0). There is no significant difference between the two groups.

Problem 3. A high school cheerleading teacher wants to test the effects of positive reinforcement on students. For this study he matches 10 cheerleaders on their ability to perform certain cheers. One cheerleader from each matched pair is assigned at random to one of two cheerleading teachers. For 50 hours each cheerleader is taught new routines. One teacher uses positive reinforcement, while the other continues his same program. At the end the students are tested and rated from 1 to 15 on their ability to perform. The results are expressed in the following table.

Pair	Positive Group	Regular Group
A	15	5
B	9	10
C	8	1
D	14	2
E	9	4
F	7	2
G	13	6
H	11	4
I	14	14
J	12	5

Level of significance is .05.

Null hypothesis: there is no significant difference between the positively reinforced group and the regular instructed group.

Alternative hypothesis states there is a difference between the two groups. This is a two-tail test.

Answer. T = 1. The table value is 6. (Note the N value is 9 because of the 0 for one pair of N's.) We reject the null hypothesis, since the T value (1) is less than the table value (6). There is a significant difference between the two groups.

When N is Larger than 25. When N is more than 25 after you find T, you have to use another formula which is

$$z = \frac{T - \frac{N(N+1)}{4}}{\sqrt{\frac{N(N+1)(2N+1)}{24}}}$$

To find the table value you go to Table D. You find the table value by first finding your z value that you computed in the z column. For example, let's say you have a z value of 3.3, a level of significance of .05 and a two-tail test. You run your fingers down the z column till you come to 3.3, you then go across this row and under .00 you find a probability of .0005. For a two-tail test you must double this probability. Therefore, .0005 becomes .0010. If this probability .0010 is *equal to or less* than your level of significance .05 you reject the null hypothesis. In this case you reject.

KRUSKAL-WALLIS TEST

Requirements

1. Data is at the ordinal level.
2. Three or more independent groups.
3. Simple random sample.

☐ **Formula: Kruskal-Wallis Test**

$$H = \frac{12}{N(N+1)} \left[\frac{(\Sigma R_1)^2}{n_1} + \frac{(\Sigma R_2)^2}{n_2} + \frac{(\Sigma R_3)^2}{n_3} + \cdots + \frac{(\Sigma R_k)^2}{n_k} \right] - 3(N+1)$$

Example. A consumer panel is rating the frozen turkey of three frozen food companies on a scale of 1 to 60. They want to find out whether or not the turkey of these companies differs in overall quality. The table below gives the ratings for the three companies.[*]

Scores Brand A	R_1	Scores Brand B	R_2	Scores Brand C	R_3
24	3	60	9	20	2
30	4	50	8	45	7
35	5	40	6	15	1

Instructions: When you rank the scores, combine all the scores together. You will put your answers in Boxes A, B, or C, then in the summary table. Later you will transfer these answers to the Kruskal-Wallis formula.

Level of significance is set at .05.

Null hypothesis states there is no significant difference among the products of the three companies.

Alternative hypothesis states there is a difference.

---

[*] The ranks are given. You can check to see if you rank properly by using the formula $\Sigma R_1 + \Sigma R_2 + \Sigma R_3 = N[(N+1)/2]$. N is defined as the number of subjects. For this problem the scores have been ranked.

Procedure	Record	Press	Display

$$\boxed{A} = \frac{12}{N(N+1)}$$

1. Find N. Count the number of individuals in all the groups. Record the answer opposite N = ___ . N=9

2. Find $N(N+1)$: Add 1 to N then multiply the answer by N. Put the answer in memory.

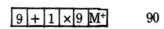   90

3. Find $\frac{12}{N(N+1)}$. Divide 12 by memory and write the answer in Box A.

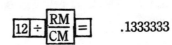   .1333333

$$\boxed{A = .1333333} = \frac{12}{N(N+1)}$$

Now transfer the answer (.1333333) to the summary table under Box A.

Procedure	Record	Press	Display
4. Clear memory and display.		RM/CM RM/CM C	0

$$\boxed{B} = \left[(\Sigma R_1)^2 + \frac{(\Sigma R_2)^2}{N_2} + \right.$$

Procedure	Record	Press	Display

$$\frac{(\Sigma R_3)^2}{N_3} + \ldots + \frac{(\Sigma R_k)^2}{n_k}\Bigg]$$

1. Find $\frac{(\Sigma R_1)^2}{n_1}$

   Add the ranks for Group A, square the sum then divide by the number of individuals in Group A, and put the answer in memory plus to be added.

   Press: `3 + 4 + 5 × = ÷ 3 M+`   Display: 48

2. Find $\frac{(\Sigma R_2)^2}{n_2}$

   Add the ranks for Group B, square your sum, then divide by the number of individuals in Group B and put the answer in memory plus to be added.

   Press: `9 + 8 + 6 × = ÷ 3 M+`   Display: 176.33333

3. Find $\frac{(\Sigma R_3)^2}{n_3}$.

   Press: `2 + 7 + 1 × = ÷ 3 M+`   Display: 33.33333

Procedure	Press	Display
Add the ranks for Group C, Square your sum, then divide by the number of individuals in Group C and put the answer in memory plus to be added.		
4. Find $\left[\dfrac{(\Sigma R_1)^2}{n_1} + \dfrac{(\Sigma R_2)^2}{n_2} + \dfrac{(\Sigma R_3)^2}{n_3} + \ldots + \dfrac{(\Sigma R_k)^2}{n_k}\right]$. Recall memory and write this answer in Box B.	$\boxed{\dfrac{RM}{CM}}$	257.66666

$$\boxed{B = 257.66666} = \left[\dfrac{(\Sigma R_1)^2}{n_1} + \dfrac{(\Sigma R_2)^2}{n_2} + \dfrac{(\Sigma R_3)^2}{n_3} + \ldots + \dfrac{(\Sigma R_k)^2}{n_k}\right]$$

Now transfer the answer 257.66666 to the summary table under Box B.

Procedure	Press	Display
$\boxed{C} = 3(N+1)$		
1. Add 1 to N (refer to A_1) then multiply the results by 3.	$\boxed{9}\boxed{+}\boxed{1}\boxed{\times}\boxed{3}\boxed{=}$	30

Procedure	Press	Display

2. Put the answer in Box C.

$$\boxed{C = 30} = 3(N + 1)$$

Now transfer the answer (30) to the summary table under Box C.

Instructions: We will now substitute in the formula

$$H = \frac{12}{N(N+1)}\left[\frac{(\Sigma R_1)^2}{n_1} + \frac{(\Sigma R_2)^2}{n_2} + \frac{(\Sigma R_3)^2}{n_3} + \cdots + \frac{(\Sigma R_k)^2}{n_k}\right] - 3(N+1)$$

the numbers from the summary table.

Summary Table

A	B	C
.1333333	257.66666	30

$H = A[B] - C$
$H = .1333333\,[257.66666] - 30$

Procedure	Press	Display
$\boxed{H}$ Find H. Multiply .1333333 by memory and subtract 30 from your answer. The display is H. $H = 4.35546$	$\boxed{.1333333} \times \boxed{\frac{RM}{CM}} - \boxed{30} =$	4.35546

Instructions: For the majority of cases you will consult Table A, the chi square distribution, to find your table value. Use a df = k − 1. Reject the null hypothesis if your value for H is equal to or greater than your table value. However, when your sample sizes are 5 or less and there are only 3 groups, you use Table H the tables Kruskal-Wallis devised. You reject the null hypothesis if your value for H has a probability of occurring that is equal to or less than your level of significance.

Statistical Decision. Write your answers in the blanks to the left

H = <u>4.36</u> 1. Record your Kruskal-Wallis value rounded to the hundredths place.

p = <u>.100</u> 2. Find the right group size for n_1, n_2, n_3 in the sample size column in Table H. In this case 3, 3, 3. Run your finger across the row into the next column which is H. You next go down the H column until you find a value closest to the H value you computed (4.36) trying not to exceed it. The value is 4.6222. Next, look at the number directly next to it in the probability column. The probability of this H occurring is .100. Since your H is 4.36, it occurs more than .100 of the time. Record this probability in the blank to the left.

<u>Fail to reject</u> 3. If the probability is equal to or less than your level of significance (.05) reject the null hypothesis. Write fail to reject in the blank to the left.

For this problem the frozen-food companies are the same.*

Problem 1. Bob, a school principal, wants to find out if there is a difference between his Spanish teachers. He randomly selects four groups of children and assigns each to a different Spanish teacher. After three months, he has a panel of language experts rate each child on his Spanish proficiency. They are rated on a scale of 1 to 37. Scores for each child are shown in the following table.

Scores Group A	R_1	Scores Group B	R_2	Scores Group C	R_3	Scores Group D	R_4
26	6	33	10	28	7	35	11
32	9	25	5	37	12	31	8
10	2	14	4	13	3	9	1

Level of significance is .01.
Null hypothesis states there is no significant difference among the four groups.
Alternative hypothesis states there is a difference.

Answer. H = .33. Use Table A with a df = k − 1. (df = 3). You have a table value of 11.34. You fail to reject the null

---
*Note the Kruskal-Wallis test only tells you whether the groups are different or the same. For you to locate *specific* differences see the multiple comparisons procedure designed by Ryan, T.A. "Significance Tests for Multiple Comparison of Proportions, Variances and Other Statistics." Psychological Bulletin 57, 4 (1960): 318-328.

hypothesis because your value for H is not greater than your table value. There appears to be no significant difference between the four groups.

Problem 2. A consumer advocate group was rating three different blue jean companies. These companies are rated on the degree to which their blue jeans are wrinkle free after one washing. They are rated 1 (poor) to 20 (excellent). The results are shown in the accompanying table.

Company A Scores	R_1	Company B Scores	R_2	Company C Scores	R_3
1	1	15	10	16	11
2	2	19	14	14	9
5	3.5	18	13	17	12
5	3.5	20	15	12	8
6	5	10	7	8	6

Level of significance is .05.

Null hypothesis states there is no significant difference among the three groups.

Alternative hypothesis states there is a difference.

Answer. H = 10.22. Use Table H. You enter the table with the column size 5, 5, 5. The probability of 10.22 occurring is less than .009. Since the probability is less than your level of significance (.05), reject the null hypothesis. There was a significant difference in the way the blue jeans were rated. The companies' products are not the same.

Problem 3. An agronomist was interested in doing research on corn growing. He wanted to test four different methods of cultivation. He had a panel of experts rate the quality of the corn produced by these four different methods. They were rated 1 (poor) to 20 (excellent). The results are shown in the accompanying table.

Method A Scores	R_1	Method B Scores	R_2	Method C Scores	R_3	Method D Scores	R_4
1	1	16	14	14	13	5	5
3	3	10	9	11	10.5	8	8
4	4	11	10.5	13	12	2	2
6	6	19	15	20	16	7	7

Level of significance is .01

Null hypothesis states there is no significant difference among the four groups.

Alternative hypothesis states there is a difference.

Answer. H = 11.70. Use Table A with a df = k − 1. (df = 3). You have a table value of 11.34. You reject the null hypothesis because your value for H(11.70) is greater than your table value (11.34). The four groups are not the same.

FRIEDMAN TEST

Requirements

1. The data is ordinal.
2. Three or more related groups.
3. The sample is drawn at random from matched scores.

☐ **Formula: Friedman Test**

$$\chi_r^2 = \frac{12}{Nk(k+1)} [\Sigma R_i^2] - 3N(k+1)$$

Example. Nine subjects formed three blocks of three matched subjects per block according to scores they were given on a psychological inventory on aggression. The three subjects in each block were then randomly assigned to three experimental treatments; each gave them a different type of aggression training. All the subjects were then placed in the same role-playing situation and rated in aggression by judges who were unaware of their previous training. The results have been recorded in the table below.

Block	Group I	Group II	Group III
A	27	25	30
B	16	10	15
C	20	24	25

Instructions: Rank the scores for *each row separately*. Record the results in a table. This ranking has been done for you*.

Block	Group I R_1	Group II R_2	Group III R_3
A	2	1	3
B	3	1	2
C	1	2	3

You will put your answers in Box A, Box B, or Box C, then in the summary table. Later you will transfer these answers to the χ_r^2 formula.

Level of significance is set at .01.

Null hypothesis states there is no significant difference in the

---

*Check your ranking by seeing if the sum of the ranks ($\Sigma R^1 + \Sigma R^2 + \Sigma R^3$ = 1/2 Nk (k + 1). N in this case is defined as number of rows and k is defined as the number of columns.

overall ratings of the three groups.

Alternative hypothesis states that the ratings of the four groups are different.

Procedure	*Record*	*Press*	*Display*

$$\boxed{A} = \frac{12}{Nk(k+1)}$$

1. Find N. Count the number of rows; write the answer opposite $N = \underline{}$. $N = 3$

2. Find k. Count the number of columns and write the answer opposite $k = \underline{}$. $k = 3$

3. Find $\frac{12}{Nk(k+1)}$.

   a. Add one to K (refer to A.2) $\boxed{3}\boxed{+}\boxed{1}\boxed{=}$ 4

   b. Multiply the answer times N (refer to A.1) then times k, and store it in memory. $\boxed{\times}\boxed{3}\boxed{\times}\boxed{3}\boxed{M^+}$ 36

   c. Divide 12 by memory and write the answer in Box A. $\boxed{12}\boxed{\div}\boxed{\frac{RM}{CM}}\boxed{=}$ .3333333

4. Clear memory and display $\boxed{\frac{RM}{CM}}\boxed{\frac{RM}{CM}}\boxed{C}$ 0

$$\boxed{A = .3333333} = \frac{12}{Nk(k+1)}$$

Now transfer the answer .3333333 to the summary table under Box A.

Procedure	Record	Press	Display
$\boxed{B} = [\Sigma R_i^2]$			
1. Find $(\Sigma R_1)^2$. Find the sum of the R_1 column, square this sum and put it in memory plus to be added.		$\boxed{2}\boxed{+}\boxed{3}\boxed{+}\boxed{1}\boxed{\times}\boxed{=}\boxed{M^+}$	36
2. Find $(\Sigma R_2)^2$. Find the sum of R_2 columns, square this sum and put it in memory plus to be added to the preceding number.		$\boxed{1}\boxed{+}\boxed{1}\boxed{+}\boxed{2}\boxed{\times}\boxed{=}\boxed{M^+}$	16
3. Find $(\Sigma R_3)^2$. Find the sum of the R_3 column, square this sum and put it in memory plus to be added to the preceding number.		$\boxed{3}\boxed{+}\boxed{2}\boxed{+}\boxed{3}\boxed{\times}\boxed{=}\boxed{M^+}$	64
4. Recall memory plus and write the answer in display in Box B.		$\boxed{\dfrac{RM}{CM}}$	116

$$\boxed{B = 116} = [\Sigma R_i]^2$$

Now transfer the answer (116) to the summary table under Box B.

Procedure	Press	Display
$\boxed{C} = 3N(k+1)$		
1. Add one to k (refer to A.2)	$\boxed{3}\boxed{+}\boxed{1}\boxed{=}$	4
2. Multiply the display by N (refer to A.1) then times this value times 3.	$\boxed{\times}\boxed{3}\boxed{\times}\boxed{3}\boxed{=}$	36
3. Write the answer for $3N(k+1)$ in Box C.		

$$\boxed{C = 36} = 3N(k+1)$$

Now transfer the answer (36) to the summary table under Box C.

Instructions: We will now substitute in the formula

$$\chi r^2 = \frac{12}{Nk(k+1)} [\Sigma R_i^2] - 3N(k+1)$$

the numbers from the summary table.

Summary Table

A	B	C
.3333333	116	36

$\chi r^2 = [A][B] - C$
$\chi r^2 = [.3333333][116] - 36$

Procedure	Record	Press	Display
Find χr^2.			
1. Multiply .3333333 (Box A) times memory (Box B), then subtract 36 (Box C).	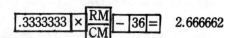		2.666662

| Procedure | Record | Press | Display |

2. $\chi r^2 =$ 2.666662. Write this answer opposite $\chi r^2 =$ ___ in the record column.

$\chi r^2 = 2.666662$

Instruction: If you have three columns with two to nine rows or four columns with two to four rows, consult Table H. For all other situations use Table A, the chi square distribution.

Statistical Decision. Write your answers in the blanks to the left.

$\chi r^2 = \underline{2.67}$ 1. Record your Friedman value rounded to the hundredths place.

$p = \underline{.361}$ 2. Find the probability using Table H. Find the correct column size—k = 3, in this case. Next find the correct row size—N = 3. You go next to the χr^2 column and run your fingers down this column until you find a value that comes closest to the χr^2 value you computed (2.67) trying not to exceed it. The value is 2.667. Next, look at the number directly next to it in the probability column. The probability of χr^2 occurring is .361. Record this probability in the blank to the left.

<u>Fail to reject</u>   3. If the probability you found is equal to or less than your level of significance (.05) reject the null hypothesis. Write fail to reject. The results shows there is no difference among the ratings of the four different types of training.*

Problem 1. A researcher wanted to know if dress influenced the way students would rate a professor on the first day of class. Four political science professors prepared the same introductory speech to be read in the same monotone to their political science class. The students were matched ahead of time on the basis of socioeconomic level and grades, and the members of each matched

---

*The comments concerning the Kruskal-Wallis test apply here. If there was a significant difference you use a significance test for multiple comparisons to locate specific differences. A discussion of the theory underlying these multiple comparison tests is found in Wilcoxon, F. and Wilcox, R. *Some Rapid Approximate Statistical Procedures*, Pearl River, New York Lederle Laboratories, 1964, pp. 9-12.

quadruplet were randomly assigned to a classroom. The students were then asked to fill out a questionnaire on their professor. The results are shown below.

Level of significance: .01

Null hypothesis states there is no significant difference between the classes.

Alternative hypothesis states that the groups are different.

Student Rater	Class I	Class II	Class III	Class IV
A	35	44	66	22
B	64	23	35	56
C	12	10	9	8
D	3	5	6	7
E	50	60	70	80
F	82	75	73	60
G	11	14	20	1
H	4	6	2	10
I	27	35	77	78
J	5	9	16	19

Instructions: Remember rank the scores for each row separately. Use Table A with a df = k − 1.

Answer. $\chi r^2 = 1.08$. The table value is 11.34. You use Table A with a df = k − 1. Since your χr^2 value (1.08) is not equal to or greater than your table value (11.34) we fail to reject the null hypothesis. There is no significant difference among the four groups.

Problem 2. Carlos was in charge of ordering running shoes for his jogging camp. He wanted to know if there was any difference in the quality of the running shoes made by three different companies. He matched runners ahead of time on their weight, height and running ability, and a member of each matched triplet was randomly assigned to the shoes of a different company. He then had the subjects run 10 miles a day for one month using the same running shoes. At the end of this time interval, the subjects' shoes were rated by a panel of experts. The highest score attainable was a 20.

Matched Triplet	Company X	Company Y	Company Z
A	15	20	10
B	18	17	12
C	19	10	15

Matched Triplet	Company X	Company Y	Company Z
D	8	12	6
E	10	9	12
F	17	11	7
G	19	8	9

Level of significance: .05.

Null hypothesis states that there is no significant difference among the companies.

Alternative hypothesis states that the companies are different.

Answer. $\chi r^2 = 3.71$. The table used is Table H. The probability of 3.71 occurring is .192. Since this probability is not equal to or less than the level of significance (.05), we fail to reject the null hypothesis. There is no significant difference among the three companies.

Problem 3. A music director wants to know if there is any difference in the three voice training programs being used at his institute. He matches singers on their ability to perform certain exercises. He then assigns a member of each matched triplet to Group A, another to Group B, and third to Group C. Group A uses training program one, Group B uses training program two, and Group C uses training program three. At the end of six months, the singers are rated on their performance.

Level of significance is set at .01.

Null hypothesis states there is no significant difference in the overall ratings of the three groups.

Alternative hypothesis states that the ratings of the three groups are different.

Matched Triplet	Program I	Program II	Program III
A	25	30	25
B	29	26	20
C	35	33	31
D	15	18	20
E	10	23	21
F	10	20	22
G	15	18	6
H	10	12	4
I	22	23	24
J	19	16	20

Answer. $\chi r^2 = 1.55$. Use Table A, the chi square distribution; df = k − 1 (df = 2). Since your χr^2 value (1.55) is not equal to or greater than your table value (9.21) we fail to reject the null hypothesis. There is no significant difference among the three groups.

TEST 6

Directions: Fill in the blanks with the names of the appropriate tests. Read each description carefully.

1. Ordinal level of measurement, five related groups: _____
2. Nominal level of measurement, two independent groups: _____
3. Ordinal level of measurement, one group: _____
4. Nominal level of measurement, three related groups: _____
5. Ordinal level of measurement, four independent groups: _____
6. Ordinal level of measurement, two independent groups: _____

ANSWERS

1. Friedman test
2. Chi square test (II)
3. Kolmogorov-Smirnov test
4. Cochran Q test
5. Kruskal-Wallis test
6. Mann-Whitney U test

Chapter 7
Interval Tests

T-TEST(I)

Requirements
1. The level of measurement is at least interval.
2. The sample is drawn from a population that has a normal distribution.
3. It is a random sample.
4. The population mean is known and the population standard deviation unknown.

☐ **Formula: t-Test(I)**

$$t = \frac{\overline{X}-\mu}{\sqrt{\dfrac{\Sigma X^2 - \dfrac{(\Sigma X)^2}{N}}{N(N-1)}}}$$

Example. A flying academy is concerned about its pilots' knowledge. Last year the mean score on the flying test was shown to be 6. The standard deviation for the scores is unknown. The flying academy wants to test the hypothesis that the average score for this present class on the test is different.

Level of significance: .05.

Null hypothesis states the mean of the population is 6.

Alternative hypothesis states that the mean of the population is *not* 6.

A two-tail test is used.

Students' Scores
4, 5, 6, 7, 8, 9, 10, 11, 12, 10

Instructions: You will put your answers in Box A, Box B, Box C or Box D, then in the summary table. Later, you will transfer these answers to the t(I) formula. You are given the population mean which is 6. $\mu=6$.

Procedure	Record	Press	Display
$\boxed{A}=\overline{X}-\mu$			
1. Find N. Count the number of scores and write the answer opposite N=___.	N = 10		
2. Find ΣX. Add the scores and write the answer opposite ΣX=___.	ΣX=82	$\boxed{4}+\boxed{5}+\boxed{6}+\boxed{7}+\boxed{8}+\boxed{9}+$ $\boxed{10}+\boxed{11}+\boxed{12}+\boxed{10}=$	82
3. Find $\overline{X}$. Divide the display by N (refer to A.1).		$\boxed{\div}\;\boxed{10}\;\boxed{=}$	8.2
4. Find $\overline{X}-\mu$. Subtract the population mean (refer to instructions) from the display and write the answer in Box A.		$\boxed{-}\;\boxed{6}\;\boxed{=}$	2.2

$$\boxed{A = 2.2} = \overline{X}-\mu$$

Now transfer the answer (2.2) to the summary table under Box A.

Procedure	Record	Press	Display
$\boxed{B}=\Sigma X^2$			
1. Square each score then		$\boxed{4}\times\boxed{4}\;\boxed{M^+}$ $\boxed{5}\times\boxed{5}\;\boxed{M^+}$	16 25

Procedure	Record	Press	Display
find the sum of the squares.		6 × 6 M+	36
		7 × 7 M+	49
		8 × 8 M+	64
		9 × 9 M+	81
		10 × 10 M+	100
		11 × 11 M+	121
		12 × 12 M+	144
		10 × 10 M+	100
2. Recall from memory the ΣX^2. Write the number in display (736) in Box B.		RM/CM	736

$$\boxed{B = 736} = \Sigma X^2$$

Now transfer the answer (736) to the summary table under Box B.

Procedure	Record	Press	Display
Clear the memory and display. $\boxed{C} = \frac{(\Sigma X)^2}{N}$		RM/CM C	0
1. Find ΣX (refer to A.2). Next square this number.		82 × =	6724
2. Divide the display by N (refer to A.1) and write the answer 672.4 in Box C.		÷ 10 =	672.4

$$\boxed{C = 672.4} = \frac{(\Sigma X)^2}{N}$$

Now transfer the answer (672.4) to the summary table under Box C.

Procedure	Record	Press	Display
$\boxed{D} = N(N-1)$			
Find $N(N-1)$			
1. Subtract one from N (refer to A.1) then multiply the answer by N and write the answer in display in Box D.		10 − 1 × 10 =	90

$$\boxed{D = 90} = N(N-1)$$

Now transfer the answer (90) to the summary table under Box D.

Instructions: We will now substitute in the formula

$$t = \frac{\overline{X} - \mu}{\sqrt{\dfrac{\Sigma X^2 - \dfrac{(\Sigma X)^2}{N}}{N(N-1)}}}$$

the numbers from the summary table.

Summary Table

A	B	C	D
2.2	736	672.4	90

$$t = \frac{A}{\sqrt{\dfrac{B-C}{D}}} = \frac{2.2}{\sqrt{\dfrac{736-672.4}{90}}}$$

Procedure	Record	Press	Display
Find t(I).			
1. Subtract C from B and divide the display by 90 (Box D).		736 − 672.4 ÷ 90 =	.7066666
2. Square root the display and put the answer in memory.		√ M⁺	.8406346

103

Procedure	Record	Press	Display
3. Divide 2.2 (Box A) by the number in memory.		2.2 ÷ RM/CM =	2.6170704
4.	t(I) 2.617074		

Statistical Decision. Write your answer in the blank to the left.

t(I) = <u>2.62</u>   1. Record your value for t(I) in the blank to the left. Round the value to the hundredths place.

df <u>9</u>   2. Find the df. Use the Formula N−1. Subtract 1 from the number of scores. Write the answer (9) in the blank to the left.

Table Value <u>2.262</u>   3. Find the table value. Use the t-distribution, Table I. Enter the table at your df, run your fingers across this row until you are under your predetermined level of significance (.05), and proper tail (two-tail). Record the value found in the blank to the left.

<u>reject</u>   4. If the value you found for t(2.62) is equal to or greater than the table value (2.262) reject the null hypothesis. Write reject.

There is a difference between the mean of this group of flyers and the assumed population mean of 6.

Problem 1. Mr. Clancey wants to test the hypothesis that the mean physics score for a population of high school students is 10. The standard deviation for the scores is unknown. He draws a random sample of nine students from the population and records their scores. The results are shown below.

Level of significance is .01.
Null hypothesis: the mean of the population is 10.
Alternative hypothesis: the mean of the population is greater than 10. Use a one-tail test.

Students' Scores
11 8 14 25 28 4 3 2 9

Answer. t(I) = .50. The table value is 2.896. We fail to reject the null hypothesis. The t value (.50) is not equal to or greater than the table value (2.896). The mean of the sample is not significantly different from the assumed population mean of 10.

Problem 2. A manager of a computer company has been dealing with X company for three years, but she is thinking about switching to Y computers because they are cheaper. Y salesmen claim their product's repair record is the best. Experience over the years has shown X computers, with constant use, have a trouble-free mean running time of 1000 hours. A random sample of 10 Y computers were tested and their scores are shown.

The *level of significance* is .01.

The *null hypothesis* states that the sample mean is smaller than 1000.

The *alternative hypothesis* states that the sample mean is smaller than 1,000. A one-tail test is used.

Y computer scores

900, 850, 400, 333, 500, 600, 700, 777, 330, 550

Answer. $t(I) = -6.17$. The table value is 2.821. We reject the null hypothesis. The t value (6.17) is greater than the table value (2.821). The sample mean is significantly smaller than 1000.

Problem 3. Ms. Wong is tired of hearing how much gas mileage different cars get. She is told by a salesman that car "T" gets 27 miles to a gallon. She tests this hypothesis by running a sample of eleven T cars over a 200 mile course.

The *level of significance* is .05.

The *null hypothesis* states the population mean equals 27.

The *alternative hypothesis* states the population mean is not equal to 27. A two-tail test is used.

T Cars

15, 20, 26, 27, 20, 21, 23, 25, 28, 18, 17

Answer. $t(I) = -3.399$. The table value is 2.228. We reject the null hypothesis. The t value (−3.99) is greater than the table value (2.228). The mean of the sample is significantly different from the assumed population mean of 27. The cars do not get 27 miles to a gallon.

T-TEST (II)

Requirements

1. The samples are drawn at random.
2. Interval level of measurement.
3. The populations are both normally distributed.

4. The populations have the same variances.*
5. The two groups are independent.

☐ **Formula: t/test(II)**

$$\frac{\overline{X}_1 - \overline{X}_2}{\sqrt{\frac{(\Sigma X_1)^2 - \frac{\Sigma X_1^2}{N_1} + \Sigma X_2^2 - \frac{(\Sigma X_2)^2}{N_2}}{N_1 + N_2 - 2} \cdot \left(\frac{N_1 + N_2}{N_1 \cdot N_2}\right)}}$$

Considering requirement 3, if there is a severe departure from normality it will have little effect on the results when the sample sizes are 30 or more. You can violate the assumption as long as your samples are not extremely small (Hay, W. L. *Statistics for the Social Sciences*, 2nd ed. New York: Holt, Rinehart and Winston, 1973).

For requirement 4, when the sample sizes are equal and the populations do not have the same variances, there is little effect on the conclusions reached by the t-test. However, this is not the case when the samples are extremely small and of unequal sizes.

Example. A principal is interested in determining the effects of two new teaching methods. He compares two groups on math achievement test. The results are recorded in the table below.

Level of significance: .05.

Null hypothesis states there is no significant difference between the two groups.

Alternative hypothesis states that the two groups are different. A two-tail test is used.

Achievement Scores

Group I	Group II
4	4
5	9
6	2
7	2
8	1
9	3
3	4
4	2
	3
	2

---

*Variance is the standard deviation squared. If the standard deviations are nearly equal then we can say the variances are equal.

Instructions: You will put your answers in Box A, Box B, Box C, Box D and Box E, then in the summary table. Later, you will transfer these answers to the t-Test(II) formula.

Procedure	Record	Press	Display
$\boxed{A} = \overline{X}_1 - \overline{X}_2$			
1. Find ΣX_1. Add the scores for Group I. Write the answer opposite the $\Sigma X_1 = \_\_\_$ in the record column.	$\Sigma X_1 = 46$	`4 + 5 + 6 + 7 + 8 + 9 + 3 + 4 =`	46
2. Find N_1. Count the number of scores in Group I. Write the answer opposite $N_1 = \_\_\_$ in the record column.	$N_1 = 8$		
3. Find $\overline{X}_1$. Divide ΣX_1 (record column) by N_1 (record column) then put the answer in memory.		`46 ÷ 8 M+`	5.75
4. Find ΣX_2. Add the scores for Group II. Write the answer opposite the $\Sigma_2 = \_\_\_$		`4 + 9 + 2 + 2 + 1 + 3 + 4 + 2 + 3 + 2 =`	32

Procedure	Record	Press	Display
in the record column.	$\Sigma X_2 = 32$		
5. Find N_2. Count the number of scores in Group II. Write the answer opposite $N_2 = \underline{}$ in the record column.	$N_2 = 10$		
6. Find $\overline{X}_2$. Divide ΣX_2 (display) by N_2 (record column) write the answer in the record column opposite $\overline{X}_2 = \underline{}$	$\overline{X}_2 = 3.2$	÷ 10 =	3.2
7. Find $\overline{X}_1 - \overline{X}_2$. Subtract the mean from Group II (A.6) from the mean from Group I (memory). Write the answer in Box A and clear memory and display.		RM/CM − 3.2 =	2.55
		RM/CM RM/CM C	0

$$\boxed{A = 2.55} = \overline{X}_1 - \overline{X}_2$$

Now transfer the answer (2.55) to the summary table under Box A.

Procedure	Record	Press	Display

$\boxed{B} = \Sigma X_1^2 - \dfrac{(\Sigma X_1)^2}{N_1}$

1. Find ΣX_1^2. Square each score for Group I, then find the sum of the squares. The answer is stored in memory.

Press	Display
4 × 4 M⁺	16
5 × 5 M⁺	25
6 × 6 M⁺	36
7 × 7 M⁺	49
8 × 8 M⁺	64
9 × 9 M⁺	81
3 × 3 M⁺	9
4 × 4 M⁺	16

2. Find $\dfrac{(\Sigma X_1)^2}{N_1}$ Square ΣX_1 (refer to A.1) divide this number by N_1 (refer to A.2) write the answer opposite

   $\dfrac{(\Sigma X_1)^2}{N_1} = $ _____ in the record column.

   | 46 × = ÷ 8 = | 264.5 |

   $\dfrac{(\Sigma X_1)^2}{N_1} = 264.5$

3. Find $\Sigma X_1^2 - \dfrac{(\Sigma X_1)^2}{N_1}$.

   Subtract $\dfrac{(\Sigma X_1)^2}{N_1}$

Procedure	Record	Press	Display
(refer to B.2) from memory. Write the answer in display in Box B.		[RM/CM] [−] [264.5] [=]	31.5

$$\boxed{B = 31.5} = \Sigma X_1^2 - \frac{(\Sigma X_1)^2}{N_1}.$$

Now transfer the answer (31.5) to the summary table under Box B.

Procedure	Record	Press	Display
Clear the memory and display.		[RM/CM] [RM/CM] [C]	0

$$\boxed{C} = \Sigma X_2^2 - \frac{(\Sigma X_2)^2}{N_2}$$

1. Find ΣX_2^2. Square each score for Group II then find the sum of the squares squares. The answer is stored in memory.

4	×	4	M⁺	16
9	×	9	M⁺	81
2	×	2	M⁺	4
2	×	2	M⁺	4
1	×	1	M⁺	1
3	×	3	M⁺	9
4	×	4	M⁺	16
2	×	2	M⁺	4
3	×	3	M⁺	9
2	×	2	M⁺	4

2. Find $\frac{(\Sigma X_2)^2}{N_2}$ Square ΣX_2 (refer to

    [32] [×] [=] [÷] [10] [=]   102.4

Procedure	Record	Press	Display

A.4. Divide this number by N_2 (refer to A.5). Write the answer opposite $\frac{(\Sigma X_2)^2}{N_2} =$ ___ in the record column.

$\frac{(\Sigma X_2)^2}{N_2} = 102.4$

3. Find $\Sigma X_2^2 - \frac{(\Sigma X_2)^2}{N_2}$.

   Subtract $\frac{(\Sigma X_2)^2}{N_2}$ (refer to C.2) from memory. Write the answer in Box C.

$\boxed{\frac{RM}{CM}} - \boxed{102.4} =$ 45.6

$$\boxed{C = 45.6} = \Sigma X_2^2 - \frac{(\Sigma X_2)^2}{N_2}.$$

Now transfer the answer (45.6) to the summary table under Box C.

Clear the memory and display.

$\boxed{D} = \left(\frac{N_1 + N_2}{N_1 \cdot N_2}\right)$

1. Find N_1 and N_2 (refer to A.2) and A.5).

$\boxed{\frac{RM}{CM} \ \frac{RM}{CM}} \boxed{C}$ 0

Procedure	Record	Press	Display

Add N_1 and N_2 and put the results in memory. Also write the answer opposite $N_1+N_2 = \underline{}$ in the record column.

$N_1+N_2=18$

$\boxed{8}\boxed{+}\boxed{10}\boxed{M^+}$ 18

2. Find $N_1 \cdot N_2$. Multiply N_1 (A.2) times N_2 (A.5). Write the answer opposite $N_1 \cdot N_2 = \underline{}$ in the record column.

$N_1 \cdot N_2 = 80$

$\boxed{8}\boxed{\times}\boxed{10}\boxed{=}$ 80

3. Find $\left(\dfrac{N_1+N_2}{N_1 \cdot N_2}\right)$ Divide N_1+N_2 (memory) by $N_1 \cdot N_2$ (D.2). Write the answer in display in Box D.

$\boxed{\substack{RM\\CM}}\boxed{\div}\boxed{80}\boxed{=}$ .225

$$\boxed{D = .225} = \left(\frac{N_1 + N_2}{N_1 \cdot N_2}\right)$$

Now transfer the answer (.225) to the summary table under Box D.

Procedure   *Record*   *Press*   *Display*

Clear the memory and display. $\boxed{\frac{RM}{CM}}\boxed{\frac{RM}{CM}}\boxed{C}$ 0

$\boxed{E} = N_1 + N_2 - 2$.

1. Find $N_1 + N_2 - 2$. Subtract 2 from $N_1 + N_2$. (Refer to Step D.1.) $\boxed{18}\boxed{-}\boxed{2}\boxed{=}$ 16
2. Write the answer in Box E.

$$\boxed{E = 16} = N_1 + N_2 - 2.$$

Now transfer the answer (16) to the summary table under Box E.

Instructions: We will now substitute in the formula

$$t = \frac{\bar{X}_1 - \bar{X}_2}{\sqrt{\frac{\Sigma X_1^2 - \frac{(\Sigma X_1)^2}{N_1} + \Sigma X_2^2 - \frac{(\Sigma X_2)^2}{N_2}}{N_1 + N_2 - 2} \left(\frac{N_1 + N_2}{N_1 \cdot N_2}\right)}}$$

the numbers from the summary table.

Summary Table

A	B	C	D	E
2.55	31.5	45.6	.225	16

$$t = \frac{A}{\sqrt{\frac{[B]+[C]}{E}} \cdot (D)}$$

$$t = \frac{2.55}{\sqrt{\frac{[31.5]+[45.6]}{16}} \cdot .225}$$

Procedure	Record	Press	Display

Find t-Test(II).

1. Add B + C, then divide by E.
   Press: `31.5` + `45.6` ÷ `16` = → 4.81875

2. Multiply the display by D, then find the square root and store the answer in memory.
   Press: × `.225` = √ M⁺ → 1.0412582

3. Divide A by memory.
   Press: `2.55` ÷ RM/CM = → 2.4489603

4. t-Test(II) = 2.4489603.

Statistical Decision. Write the answers in the blanks to the left.

t = <u>2.45</u>
1. Write the value for t-Test(II) in the blank to the left. First round it to the hundredths place.

df = <u>16</u>
2. Find df. Use the formula $N_1 + N_2 - 2$ to find df. Add 8 + 10 and subtract 2.

Table value = <u>2.120</u>
3. Find the table value. Use Table G. Enter at your df and run your fingers across this row until you are under your predetermined level of significance (.05) and proper tail (two-tail). Write 2.120 in the blank.

<u>Reject</u>
4. If the absolute value you found for t(2.45) is equal to or greater than the table value (2.120), reject the null hypothesis. Write reject at the left.

There appears to be a significant difference between the two groups on math achievement.

Problem 1. A psychologist who is working with autistic children wants to determine the effects of two innovative training techniques. He compares two independent groups on a vocabulary achievement test.

Level of significance: .01.

Null hypothesis states there is no significant difference between the two groups.

Alternative hypothesis states that Group I will do better than Group II. Use a one-tail test.

Vocabulary Scores

Group I	Group II
3	1
4	2
5	3
7	5
8	4
10	3
12	

Answer. t-Test (II) = 2.77. The table value is 2.718. We reject the null hypothesis, since the t value (2.77) is greater than the table value (2.718). There is a significant difference between the two training techniques.

Problem 2. An owner of an accounting firm is thinking of switching to another, less-expensive calculator firm. This firm claims they have a new battery that will outperform any on the market and make their calculator the most practical to use. One sample of calculators is randomly selected from the old firm and one sample of calculators is randomly selected from the new firm. They both are tested and their scores recorded.

Level of significance: .05.

Null hypothesis states there is no significant difference between the groups of batteries.

Alternative hypothesis states that the two groups are different. A two-tail test is used.

Battery Life Scores

Old Company	New Company
80	70
70	80
50	90
40	80
60	100
50	80
45	88
58	77
76	98
	86

Answer. t(II) test = -4.82. The table value is 3.110. We reject the null hypothesis, since the t value (-4.82) is greater than the table value (3.110). There is a significant difference between the two companies batteries.

Problem 3. Two randomly selected groups of sociology students took a sociology test. One group was taught by the programmed learning approach, while the other group had been taught by the lecture approach. In the table below the scores of the members of the two independent groups are given.

Level of significance is .01.

Null hypothesis states there is no significant difference between the two groups of students.

Alternative hypothesis states that the two groups are different. Use a two-tail test.

Programmed Group	Lecture Group
20	10
19	9
18	8
10	7
12	5
10	10
15	5
	4

Answer. t(II) test = 4.35. The table value is 3.012. We reject the null hypothesis, since the t value (4.35) is greater than the table value (3.012). There is a significant difference between the two teaching approaches.

T-TEST(III)

Requirements

1. The two groups are related.

2. Interval level of measurement.
3. Populations are both normally distributed.
4. Populations have the same variances.
5. Samples are randomly drawn.

☐ **Formula: t-Test(III)**

$$t = \frac{\overline{D}}{\sqrt{\frac{\Sigma D^2 - \frac{(\Sigma D)^2}{N}}{N(N-1)}}}$$

Example. The experimenter is interested in improving a child's score on the reading achievement test. Students were matched on IQ and reading achievement scores. One student from each pair is randomly assigned to either a special treatment or a control group. After seven weeks of training, another form of the reading achievement test is given to determine the effects of the program. The score for this second test is recorded in the table below.

Level of significance: .05.

Null hypothesis: there is no significant difference between the two groups.

Alternative hypothesis: the two groups are different. Use a two-tail test.

Reading Achievement Scores

Pair	Group I	Group II
A	9	10
B	15	8
C	20	7
D	14	6
E	13	5
F	10	4
G	3	12
H	10	2

Instructions: You will put your answers in Box A, Box B and Box C, then the summary table. Later you will transfer these answers to the t (III) formula.

Procedure	Record	Press	Display

$\boxed{A} = \overline{D}$

1. Find N. Count the number of pairs of scores and write the answer in the record column opposite N = ____.   N = 8

2. Find ΣD. Form the D column. Find the difference between each pair of scores and write down with its sign under a column labeled D in the record column.

   D
   −1
   7
   13
   8
   8
   6
   −9
   8

$\boxed{9}$ − $\boxed{10}$ =	−1
$\boxed{15}$ − $\boxed{8}$ =	7
$\boxed{20}$ − $\boxed{7}$ =	13
$\boxed{14}$ − $\boxed{6}$ =	8
$\boxed{13}$ − $\boxed{5}$ =	8
$\boxed{10}$ − $\boxed{4}$ =	6
$\boxed{3}$ − $\boxed{12}$ =	−9
$\boxed{10}$ − $\boxed{2}$ =	8

3. Find ΣD. You find the algebraic sum of the D column (refer to A.2 of the record column). Write the number in display opposite ΣD = ____ in the

   $\boxed{1}$ $\boxed{SC}$ + 7 + 13 + 8 +
   8 + 6 + 9 $\boxed{SC}$ + 8 =   40

Procedure	Record	Press	Display
record column.	$\Sigma D = 40$		
4. Find $\overline{D}$. Divide the number in display by N(A.1) and write the answer in Box A.		$\div$ 8 $=$	5

$$\boxed{A = 5} = \overline{D}$$

Now transfer the answer (5) to the summary table under Box A.

Procedure	Record	Press	Display
$\boxed{B} = \Sigma D^2$		1 SC $\times$ 1 SC M+	1
1. Square each score in the D column (refer to A.2), then find the sum of the squares.		7 $\times$ 7 M+	49
		13 $\times$ 13 M+	169
		8 $\times$ 8 M+	64
		8 $\times$ 8 M+	64
		6 $\times$ 6 M+	36
		9 SC $\times$ 9 SC M+	81
		8 $\times$ 8 M+	64
2. Recall from memory the answer (528) and write it in Box B.		RM/CM	528
3. Clear the memory and display.		RM/CM C	0

$$\boxed{B = 528} = \Sigma D^2$$

Now transfer the answer (528) to the summary table under Box B.

Procedure	Record	Press	Display
$\boxed{C} = \dfrac{(\Sigma D)^2}{N}$			

Procedure	Record	Press	Display
1. Find ΣD (refer to A.3) and square this value.		$\boxed{40}\boxed{\times}\boxed{=}$	1600
2. Next, divide, the display by N (refer to A.1) and write this answer (200) in Box C.	$\boxed{C = 200} = \dfrac{(\Sigma D)^2}{N}$	$\boxed{\div}\boxed{8}\boxed{=}$	200

Now transfer the answer (200) to the summary table under Box C.

Procedure	Record	Press	Display
$\boxed{D} = N(N-1)$			
1. Find N (refer to A.1) and subtract 1 from this number.		$\boxed{8}\boxed{-}\boxed{1}\boxed{=}$	7
2. Multiply the display by N and write the answer in Box D.		$\boxed{\times}\boxed{8}\boxed{=}$	56

$$\boxed{D = 56} = N(N-1)$$

Now transfer the answer (56) to the summary table under Box D.

Instructions: We will now substitute in the formula

$$t = \frac{\overline{D}}{\sqrt{\dfrac{\Sigma D^2 - \dfrac{(\Sigma D)^2}{N}}{N(N-1)}}}$$

the numbers from the summary table.

Summary Table

A	B	C	D
5	528	200	56

$$t = \frac{A}{\sqrt{\dfrac{B-C}{D}}} = \frac{5}{\sqrt{\dfrac{528-200}{56}}}$$

Procedure	Record	Press	Display
Find t-test (III)			
1. Subtract C from B.		`528` `−` `200` `=`	328
2. Divide the display by D.		`÷` `56` `=`	5.8571428
3. Square root the display and put the number in memory.		`√` `M+`	2.4201534
4. Divide A by the number in memory. t = 2.0659847		`5` `÷` `RM/CM` `=`	2.0659847

Statistical Decision. Write your answers in the blank to the left.

t = <u>2.07</u> 1. Record your value for t (III) in the blank to the left. Round the value to the hundredths place.

df = <u>7</u> 2. Find df. Use the Formula N − 1. N is the number of pairs of scores. Subtract 1 from the number of pairs of scores (8). Write 7 in the blank to the left.

Table Value = <u>2.365</u> 3. Find the table value. Use the t-distribution, Table G. Enter the table at your df, and run your fingers across this row until you are under your predetermined level of significance (.05) and proper tail (two-tail). Record the value found in the blank to the left.

<u>Fail to reject</u> 4. If the absolute value you found for t(2.07) is equal to or greater than the table value (2.365), reject the null hypothesis. Write fail to reject.

There is no significant difference between the two groups.

Problem 1. A doctor wanted to test the effect of a hormone on the growth of 20 youngsters. The youngersters were matched on sex, height, and family history. He assigned one member of each matched pair to Group A and the other member to Group B. Group A received the hormone for two years, while Group B received a placebo for two years.

Level of significance is .01.

Null hypothesis states there is no significant difference between the two groups.

Alternative hypothesis states that Group A will do better than Group B. A one-tail test is used.

Growth in Inches
Two Year Period

Pair	Group A	Group B
A	33	22
B	10	8
C	7	4
D	8	2
E	9	1
F	11	15
G	15	19
H	44	33

Answer. t(III) = 1.95. The table value is 2.998. We fail to reject the null hypothesis. The t value (1.95) is not equal to or greater than the table value (2.998). There is no significant difference between Group A and Group B.

Problem 2. Fifteen university students were given the Miller Analogy Exam. They were then asked to take a years's course on how to take this exam. At the end of the year, the same students retook the test. The results were recorded.

Level of significance is .05.

Null hypothesis states that the two groups are the same.

Alternative hypothesis states that the two groups are different. A two-tail test is used.

Miller Analogy Exam

Student	Before	After
A	40	20
B	30	40
C	60	65
D	70	65

Miller Analog Exam

Student	Before	After
E	80	82
F	72	70
G	90	95
H	96	98
I	75	77
J	60	58
K	70	65
L	88	88
M	45	46
N	60	65
O	75	76

Answer. t(III) = .04. The value is 2.145. We fail to reject the null hypothesis. The t value (.04) is not equal to or greater than the table value (2.145). There is no significant difference between the two groups.

Problem 3. Students were administered a test in physics proficiency. They were then given a lecture on relaxation during a test. The physics test was readministered after this lecture.
Level of significance is .01.
Null hypothesis states that the two groups are the same.
Alternative hypothesis states that the two groups are different. A two-tail test is used.

Physics Test Scores

Student	Before	After
A	10	15
B	15	16
C	14	13
D	13	10
E	12	14
F	15	16
G	17	18
H	20	20
I	19	20

Answer. t (III) = −1.08. The table value is 3.250. We fail to reject the null hypothesis. The t value (−1.08) is not equal to or greater than the table value (3.250). There is no significant difference between the two groups.

ONE-WAY ANALYSIS OF VARIANCE—RANDOMIZED GROUPS MODEL

Requirements
1. Three or more independent groups.
2. Individuals of the same have been randomly drawn from the population.
3. The variances are equal.
4. A normally distributed population.
5. The level of measurement is at the least interval.

☐ **Formula: One-Way Analysis of Variance Randomized Groups, F-Ratio**

$$F = \frac{MS_B}{MS_W}$$

Example. A prison warden wanted to compare the effectiveness of three different methods for teaching math to delinquent teenagers. Nine delinquent teenagers were randomly assigned to three independent groups. Each group was taught math for two months, one hour a day. At the end of this time, the delinquents were all given a math achievement test and the scores were recorded in the table below.

Level of significance is .05.

Null hypothesis states there is no significant difference between the three groups.

Alternative hypothesis states the three groups are different.

Math Achievement Scores

Group A	Group B	Group C
7	6	4
6	5	2
10	10	3

Procedure   *Record*   *Press*   *Display*

The between-group sum of squares

1. Find N. Count the number of individuals in all the groups and write the answer in

Procedure	Record	Press	Display
the record column opposite $N = \underline{}$.	$N = 9$		

2. Find
$$\frac{(\Sigma X_1)^2}{N_1} + \frac{(\Sigma X_2)^2}{N_2} + \frac{(\Sigma X_3)^2}{N_3}.$$

	Procedure	Record	Press	Display
a.	Add the Group A and record opposite $\Sigma X_1 = \underline{}$.	$\Sigma X_1 = 23$	$\boxed{7} \boxed{+} \boxed{6} \boxed{+} \boxed{10} \boxed{=}$	23
b.	Square your answer then divide by the number in that group and put the answer in memory.		$\boxed{\times} \boxed{=} \boxed{\div} \boxed{3} \boxed{M^+}$	176.33333
c.	Add the scores in Group B and record opposite $\Sigma X_2 = \underline{}$.	$\Sigma X_2 = 21$	$\boxed{6} \boxed{+} \boxed{5} \boxed{+} \boxed{10} \boxed{=}$	21
d.	Square your answer, then divide by the number in that group and put the answer in memory.		$\boxed{\times} \boxed{=} \boxed{\div} \boxed{3} \boxed{M^+}$	147

Procedure	Record	Press	Display
e. Add the scores in Group C, and record your answer opposite $\Sigma X_3 = $ ___.	$\Sigma X_3 = 9$	`4` `+` `2` `+` `3` `=`	9
f. Square your answer, then divide by the number in that group and put your answer in memory.		`×` `=` `÷` `=` `M⁺`	27

3. Find $\dfrac{(\Sigma X)^2}{N}$ the correction term.

a. Add the $\Sigma X_1 + \Sigma X_2 + \Sigma X_3$ (refer to the record column).		`23` `+` `21` `+` `9` `=`	53
b. Square the display and divide by N (refer to record column opposite 1). Write your answer in the record column and *circle* it. Label it correction term.	Correction Term (312.11111)	`×` `=` `÷` `9` `=`	312.11111

126

Procedure	Record	Press	Display

4. Find the between group sum of squares

   a. Return from memory
   $$\frac{(\Sigma X_1)^2}{N_1} + \frac{(\Sigma_2)^2}{N_2} + \frac{(\Sigma X_3)^2}{N_3}$$
   and subtract the correction term from this value. (Circled value in the record column).

	Record	Press	Display
	−312.11111	RM CM =	38.22222

   b. Record the between-group sum of squares in the summary table which follows.

   c. Find the between-group degrees of freedom $(k - 1)$. Subtract one from the number of groups you have. Record this value in the summary table which follows.

Press	Display
3 − 1 =	2

Procedure	Record	Press	Display
d. Clear memory and the display.		RM/CM RM/CM C	0

Summary Table

Source of Variation	Sum of Squares	Degrees of Freedom
Between Groups	38.22222	2

Transfer the two values in this summary table to the next one.

Procedure	Record	Press	Display

Total sum of squares

1. Square each score (see original problem) and put the value in memory.

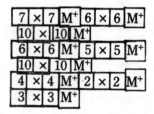

2. Recall memory and subtract the correction term (refer to the record column, circled value).

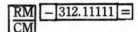

62.88889

3. Put the total sum of squares (display) in the summary table which follows.

Procedure	Record	Press	Display

4. Find the total degrees of freedom. Subtract 1 from N (record column). Put this value in the summary table which follows.

Press: `9` `-` `1` `=` Display: 8

5. Clear the memory and display.

Press: `RM/CM` `RM/CM` `C` Display: 0

Summary Table

Source of Variation	Sum of Squares	Degrees of Freedom
Between Groups	38.22222	2
Within Groups		
Total	62.88889	8

Transfer all the values in this summary table to the next one.

Procedure	Press	Display

Find F

We will now work with the following summary table and fill in the remaining values.

1. Find the within-group sum of squares. Subtract the between-group sum of squares

Press: `62.88889` `-` `38.22222` `=` Display: 24.66667

Procedure	Press	Display
from the total sum of squares.		
2. Find the within-group degrees of freedom. Subtract the between-group degrees of freedom from the total degrees of freedom.	$\boxed{8}\boxed{-}\boxed{2}\boxed{=}$	6
3. Find the between-group mean square. Divide the between-group sum of squares by the between-group degrees of freedom and put the answer in memory.	$\boxed{38.22222}\boxed{\div}\boxed{2}\boxed{M^+}$	19.11111
4. Find the within-group mean square. Divide the within-group sum of square by the within group degrees of freedom.	$\boxed{24.66667}\boxed{\div}\boxed{6}\boxed{=}$	4.111116

Procedure	Press	Display
5. Find F. Recall memory to divide the between-group mean square by the within group mean square F = 4.6486478.	[RM/CM] [÷] [4.1111116] [=]	4.6486478

Summary Table

Source of Variation	Sum of Squares	Degrees of Freedom	Mean Square	F
Between-Group	38.22222	2	19.11111	4.6486478
Within-Group	24.66667	6	4.1111116	
Total	62.88889	8		

Statistical Decision. Write your answers in the blank to the left.

F = <u>4.65</u> 1. Record your F value in the blank to the left. Round the value to the hundredths place.

Table value = <u>5.14</u> 2. Find the table value by using Table J. Run your fingers down the df column until you are at the within-group df(6), (N − k) then run your fingers across this row until you are underneath your between-group df(2) (k − 1). At this point you will have two numbers. The boldface number is for a .01 level of significance and the lightface number is for a .05 level of significance. We are interested in the .05 level of significance, therefore we record this value in the blank at the left.

<u>Fail to reject</u> 3. If the value you found for F(4.65) is equal to or greater than your table value (5.14), reject the null hypothesis. Write fail to reject.

There is no significant difference between the three groups.

Problem 1. A teacher wanted to test the effectiveness of four different programs in reinforcement on children's reading achievement. Sixteen children were randomly assigned to four independent groups. Each group was taught for eight months using its own special reinforcement program. At the end of this time period, everyone was given a reading test. The results are shown.

Level of significance is .01.

The *null hypothesis* states that there is no significant difference between the sample means.

The *alternative hypothesis* states that the four groups are different.

Reading Test Scores

Group A	Group B	Group C	Group D
20	40	25	23
25	30	20	24
15	10	12	12
10	15	17	15

Answer. $F = .43$. The table value is 27.05. The between df = 3, the within df = 12. We fail to reject the null hypothesis. The F value (.43) is not equal to or greater than your table value (27.05). There is no significant difference between the four groups.

Problem 2. A psychologist wanted to determine the effects of test anxiety on subjects taking an exam in math. Twenty subjects were randomly assigned to five independent groups. Each group was given a different dosage of test anxiety during a math test. The scores of the subjects are shown.

Level of significance is .05.

The *null hypothesis* states that there is no significant difference between the sample means.

The *Alternative hypothesis* states that the four groups are different.

Math Test Scores

Group A	Group B	Group C	Group D
80	85	86	84
70	60	65	70
60	72	70	65
90	92	94	95
97	95	94	93

Answer. $F = .03$. The table value is 8.69. The between df = 3, the within df = 16. We fail to reject the null hypothesis. The F value (.03) is not equal to or greater than your table value (8.69). There is no significant difference between the four groups.

Problem 3. A health farm owner wanted to test out four new techniques for improving physical stamina. He randomly assigned 12 individuals to four groups. Each group received a different treatment. After four weeks, they were put on a jogging machine and the time they lasted on this machine was recorded.

Level of significance is .05.

Null hypothesis: there is no significant difference between the sample means.

Alternative hypothesis states that the four groups are different.

Time on the Running Machine

Group A	Group B	Group C	Group D
2	5	10	3
4	3	15	5
5	2	9	1

Answer. $F = 10.15$. The table value is 4.07. The between df = 3, the within df = 8. We reject the null hypothesis; the groups are different. The F value (10.15) is greater than the table value (4.07).

In this problem the groups are different. The randomized groups design does not tell you which pairs of groups have significant differences. In order to find this out, we will use a significance test called the Scheffé.

SCHEFFE TEST

Instructions: The Scheffé test is to follow the randomized groups. It is used to tell you which pair of groups had differences that were significant and which pair had differences that were not significant. Referring back to practice problem number three, we will compare Group A and Group C. There are six comparisons that are possible.* We should also compare Group A versus Group B, Group A versus Group D, Group B versus Group D, and Group C versus Group D, and Group B versus Group C. The values you will need from the previous practice problem are listed in the following chart.

Summary Chart

Mean square Within	$MS_W = 4.7500012$
Group I	$n_1 = 3$
Group II	$n_2 = 3$
Between-Group Degrees of freedom	$(k - 1) = 3$
Table value	$T = 4.07$

---

*The formula $k(k - 1)/2$ tells us there are six comparisons possible taking two groups at a time.

Note that two additional values are listed in the chart n_1 and n_2. The first n_1 is the number of individuals in Group A, the second n_2 is the number of individuals in Group C. You will now put your answers in Boxes A, B, C and D. Then you will transfer these answers to the Scheffé formula.

☐ **Formula: Scheffé Test**

$$F = \frac{(\overline{X}_1 - \overline{X}_2)^2}{MS_w \; \frac{n_1+n_2}{n_1 n_2} \; (k-1)}$$

Example. Using the same level of significance as the previous practice problem (.05), find out if the differences were significant between Group A and Group C.

Group A	Group C
2	10
4	15
5	9

Procedure	Record	Press	Display

$\boxed{A} = (\overline{X}_1 - \overline{X}_2)^2$

1. Find $\overline{X}_1$. Add the scores for Group A (ΣX_1) then divide by n_1 (chart) and put the answer in memory.

   Press: $\boxed{2}\boxed{+}\boxed{4}\boxed{+}\boxed{5}\boxed{\div}\boxed{3}\boxed{M^+}$ Display: 3.6666666

2. Find $\overline{X}_2$. Add the scores for Group C (ΣX_2) then divide by n_2 (chart). Write the answer in the record

   Press: $\boxed{10}\boxed{+}\boxed{15}\boxed{+}\boxed{9}\boxed{\div}\boxed{3}\boxed{=}$ Display: 11.333333

Procedure	Record	Press	Display
column opposite $\overline{X}_2 = $ _____	$\overline{X}_2 = 11.333333$		
3. Find $(\overline{X}_1 - \overline{X}_2)^2$. Recall memory, and subtract the mean from Group II (refer to A.2) then square this difference and write the answer in Box A.		[RM/CM] − [11.333333] [×] [=]	58.777782
4. Clear the memory and the display.		[RM/CM] [RM/CM] [C]	0

$$\boxed{A = 58.777782} = (\overline{X}_1 - \overline{X}_2)^2$$

Find the Mean Square Within MS_w. Copy the value from the chart in Box B.

$$\boxed{B = 4.7500012} = MS_w$$

Procedure	Record	Press	Display
$\boxed{C} = \left(\dfrac{n_1 + n_2}{n_1 n_2}\right)$			
1. Add $n_1 + n_2$ (chart). Put the sum in memory.		[3] [+] [3] [M+]	6
2. Multiply n_1 times n_2 and write the product in the record column		[3] [×] [3] [=]	9

Procedure	Record	Press	Display

opposite
$n_1 n_2 =$ _____. $n_1 n_2 = 9$

3. Divide memory by $n_1 n_2$ (refer to C.2). Write your answer in Box C.

   Press: RM/CM ÷ 9 =   Display: .6666666

4. Clear memory and the display.

   Press: RM/CM RM/CM C   Display: 0

$$\boxed{C = .6666666} = \left(\frac{n_1 + n_2}{n_1 n_2}\right)$$

D

Find the between-group degrees freedom. Transfer the value from the chart to Box D.

$$\boxed{D = 3} = (k - 1).$$

We will now substitute and solve.

$$F = \frac{(\overline{X}_1 - \overline{X}_2)^2}{MS_W \left(\frac{n_1 + n_2}{n_1 n_2}\right)(k-1)}$$

$$= \frac{A}{(B)(C)(D)} = \frac{58.777782}{(4.7500012)(.6666666)(3)}.$$

Procedure	Press	Display

Find F

Group A vs. Group C.

1. Multiply B times C times D and put the answer in memory plus.

   Press: 4.7500012 × .6666666 × 3 M⁺   Display: 9.5000013

2. Divide A by F = 6.1871292.

   Press: 58.777782 ÷ RM/CM =   Display: 6.1871341

Statistical Decision. Write your answers in the blank to the left.

F = <u>6.19</u>   1. Record your F value in the blank to the left. Round to the hundredths place.

Table Value = <u>4.07</u>   2. Record the table value from the summary chart in the blank to the left.

<u>Significant</u>   3. If your F value (6.19) is equal to or greater than the table value (4.07), these two groups are significantly different. Write significant.

Problem. You will use the randomized group design problem and make three more comparisons. Again use the same chart values.

1. You will compare Group C and Group D and see if they are significantly different.

Group C	Group D
10	3
15	5
9	1

Answer. F = 7.31. Group C and Group D are significantly different, since the F value (7.31) is greater than the table value (4.07).

2. You will compare Group A and Group B and see if they are significantly different.

Group A	Group B
2	5
4	3
5	2

Answer. F = .01. Group A and Group B are not significantly different, since the F value (.01) is not equal to or greater than the table value (4.07).

3. You will compare Group A and Group D, and see if they are significantly different.

Group A	Group D
2	3
4	5
5	1

Answer. F = .05. Group A and Group D are not significantly different, since the F value (.05) is not equal to or greater than the table value (4.07).

ONE-WAY ANALYSIS OF VARIANCE—RANDOMIZED BLOCK MODEL
Requirements
1. Population is normally distributed.
2. Variances are equal.
3. Three or more related groups.
4. At least the interval level of measurement.
5. Subjects in each treatment are random samples from the normal population.

☐ **Formula: Randomized Block, F-Ratio**

$$F = \frac{MS_B}{MS_E}$$

Example. An experimenter wants to test the effectiveness of three different teaching techniques which claim to improve the ability to solve physics problems. Twelve students are to be used in three different treatment groups. On the basis of prior information, the experimenter finds out the subjects vary considerably in the number of problems they now can solve. The subjects are matched on the basis of a score obtained for each subject on solving physics problems in a given time frame. The subjects are now arranged in blocks. Within each block they are assigned at random one subject to each treatment. After six months of treatment, the students are tested on their problem solving ability and the results are shown in the table below.

Level of significance is .05.

Null hypothesis states there is no significant difference between the three groups.

Alternative hypothesis states the three groups are different.

Block	Treatment		
	1	2	3
1	1	3	5
2	6	7	2
3	4	4	1

Procedure	*Record*	*Press*	*Display*
1. Find N. Count the number of subjects in all the groups and			

Procedure	Record	Press	Display
write your answer opposite N=___.	N=9		
2. Find $\frac{(\Sigma X_1)^2}{n_1} + \frac{(\Sigma X_2)^2}{n_2} + \frac{(\Sigma X_3)^2}{n_3}$.			
a. Add the scores in the first column then record your answer opposite ΣX_1=___.	ΣX_1=11	[1][+][6][+][4][=]	11
b. Square your answer, then divide by the number in the column and put the answer in memory plus.		[×][=][÷][3][M⁺]	40.333333
c. Add the scores in the second column, then record your answer opposite ΣX_2=___.		[3][+][7][+][4][=]	14
d. Square your answer, then divide by the number in that column and	ΣX_2 = 14	[×][=][÷][3][M⁺]	65.333333

Procedure	Record	Press	Display
put the answer in memory plus to be added.			
e. Add the scores in the third column, then record your answer opposite $\Sigma X_3 =$ ___.	$\Sigma X_3 = 8$	$\boxed{5}\boxed{+}\boxed{2}\boxed{+}\boxed{1}\boxed{=}$	8
f. Square your answer, then divide by the number in that column and put the answer in memory plus to be added.		$\boxed{\times}\boxed{=}\boxed{\div}\boxed{3}\boxed{M^+}$	21.33333
3. Find the $\dfrac{(\Sigma X)^2}{N}$, the correction term.			
a. Add the $\Sigma X_1 + \Sigma X_2 + \Sigma X_3$ (refer to the record column opposite 2a, 2c, and 2e, in this section).		$\boxed{11}\boxed{+}\boxed{14}\boxed{+}\boxed{8}\boxed{=}$	33
b. Square the display		$\boxed{\times}\boxed{=}\boxed{\div}\boxed{9}\boxed{=}$	121

Procedure	Record	Press	Display
and divide by N (refer to the record column). Write the answer under correction term in the record column and circle it.	*Correction Term*		
4. Find the between-groups sum of squares			
a. Return from memory $\frac{(\Sigma X_1)^2}{n_1} + \frac{(\Sigma X_2)^2}{n_2} + \frac{(\Sigma X_3)^2}{n_3}$ and subtract the correction term from this value. (The circled value in the record column.) Write the answer in the summary table which follows under SS for columns.		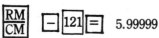	5.99999

Procedure	Record	Press	Display
b. Clear the memory and display.		[RM/CM] [RM/CM] [C]	
5. Find the between groups degrees of freedom (df). Subtract 1 from number of columns (k−1). Write this value in the Summary Table under df for the columns.		[3][−][1][=]	2

Summary Table

Source of Variation	SS	df
Between Groups	5.99999	2

Transfer the two values in this summary table to the next one.

Procedure	Record	Press	Display
1. Total sum of squares *a.* Square each score and put the answer in memory plus to be added.		[1]×[1][M+] [6]×[6][M+] [4]×[4][M+] [3]×[3][M+] [7]×[7][M+] [4]×[4][M+] [5]×[5][M+] [2]×[2][M+] [1]×[1][M+]	16 16 1
b. Recall memory and subtract the correction term (record column.)		[RM/CM] [−][121][=]	36

Procedure	Record	Press	Display
c. Put the *total sum of square* (the display) in the summary table which follows under SS for the Total			
d. Clear the memory and display.		RM/CM RM/CM C	0
2. Find the Total Degrees of Freedom. Subtract 1 from N (refer to the record column). Put this value in the summary table which follows, under df for the total.		9 − 1 =	8

Summary Table

Source of Variation	SS	df
Between-Groups	5.99999	2
Between Blocks		
Residual (Error)		
Total	36	8

Transfer all the values in this summary table to the next one.

Procedure	Record	Press	Display
1. Blocks sum of the			

| Procedure | Record | Press | Display |

squares.

a. Add the scores in the first row, square your answer, then divide by the number in the row and put that answer in memory plus.

$\boxed{1}\boxed{+}\boxed{3}\boxed{+}\boxed{5}\boxed{\times}\boxed{=}\boxed{\div}\boxed{3}\boxed{M^+}$ 27

b. Add the scores in the second row, square your answer, then divide by the number in the row and put that answer in memory plus to be added.

$\boxed{6}\boxed{+}\boxed{7}\boxed{+}\boxed{2}\boxed{\times}\boxed{=}\boxed{\div}\boxed{3}\boxed{M^+}$ 75

c. Add the scores in the third row, square your answer, then divided by the number in the row and put that answer in memory plus to be added. Find the blocks sum

$\boxed{4}\boxed{+}\boxed{4}\boxed{+}\boxed{1}\boxed{\times}\boxed{=}\boxed{\div}\boxed{3}\boxed{M^+}$ 27

Procedure	Record	Press	Display
of squares.			
d. Recall Memory and subtract the correction term from it. (Refer to the record column for this term.) Write the answer in the summary table under SS for the blocks.		RM/CM – 121 =	8
e. Clear the memory and display.		RM/CM RM/CM C	0
2. Find the blocks degree of freedom. Subtract 1 from the number of blocks (n – 1). Write the answer in the summary table under df.		3 – 1 =	2

Summary Table

Source of Variation	SS	df	MS	F
Between Groups	5.99999	2	2.999995	.5454452
Blocks	8	2	4	
Residual (Error)	22.00001	4	5.5000025	
Total	36	8		

Procedure	Press	Display
We will work with the preceding summary table and fill in the remaining values.		
1. Find the error sum of squares.		
a. Add the between-groups sum of squares to the blocks sum of squares. Put this answer in memory.	5.99999 + 8 M+	13.99999
b. Subtract the answer in memory from the total sum of squares. Place this answer in the summary table under SS for error.	36 − RM/CM =	22.00001
c. Clear memory and display.	RM/CM RM/CM C	0
2. Find the error degrees of freedom.		
a. Add the between-groups degrees of freedom and	2 + 2 M+	4

146

Procedure	Press	Display
the blocks degrees of freedom. Put the answer in memory.		
b. Subtract the answer in memory from the total degrees of freedom. Put your answer in the summary table under df for the error.	8 − RM/CM =	4
c. Clear the memory and display.	RM/CM RM/CM C	0
3. Find the between-groups mean square. Divide the between groups SS by the between groups df and put the answer in the summary table under MS for blocks.	5.99999 ÷ 2 =	2.999995
4. Find the blocks mean square. Divide the blocks SS by the blocks	8 ÷ 2 =	4

Procedure	Press	Display

df. Put the answer in the summary table under MS for the blocks.

5. Find the error mean square. Divide the error SS by the error df. Write this answer in the summary table under MS for error.

| 22.00001 ÷ 4 = | 5.5000025 |

6. The F is obtained by dividing mean square for the between-groups (MS_B) The treatment by the mean square for the error MS_e). Write this answer in the Summary table under the F.

| 2.99995 ÷ 5.5000025 = | .5454452 |

Statistical Decision. Write your answers in the blank to the left.

F = .55   1. Record your F value in the blank to the left. Round the value to the hundredths place.

Table Value = <u>6.94</u>

2. Find the table value by using Table J. Run your fingers down the df column labeled within group until you are at your error df(4), then run your fingers across this row until you are underneath your between-groups df(2). The table value is the lightface number which is for a .05 level of significance.* Write 6.94 in the blank at the left.

<u>Fail to reject</u>

3. If the value you found for F(.55) is equal to or greater than your table value (6.94), reject the null hypothesis. Write fail to reject. There is no difference between the three groups.

Problem 1. A garage owner wants to see if there is a difference in the effectiveness of four new methods for repairing tires. Twelve technicians are to be used in four different treatment groups. The subjects are matched on their ability to repair tires. They are arranged in blocks and assigned at random one subject to each method. After three months, the subjects are tested and their results measured. The scores are then recorded in a table.

Level of significance is .01.

Null hypothesis states there is no significant difference between the four groups.

Alternative hypothesis states the four groups are different in their ability to repair tires.

Tire Repair Scores

Block	Treatment Condition			
	1	2	3	4
1	1	5	4	2
2	8	3	2	3
3	5	6	4	4

Answer. F = .50. The table value is 9.78. The Between-groups df = 3, and the Residual (Error) df = 6. We fail to reject the null hypothesis. The F ratio (.50) is not equal to or greater than the table value (9.78). The four groups are not significantly different in their ability to repair tires.

Problem 2. A consumer group wants to test out three new methods for insulating refrigerators. The models are matched on their cooling volume and overall weight. The models are arranged in blocks and assigned at random one model to each method. They are then tested to see how well they save electricity in one day. The results are then recorded.

*The boldface number is for the .01 level of significance.

Power Usage for a Day

Block	Treatment Condition		
	1	2	3
1	400	450	500
2	600	625	620
3	700	675	725

Level of significance is .05.

Null hypothesis states there is no significant difference Between the three methods of insulating refrigerators.

Alternative hypothesis states the three methods are different.

Answer. F = 2.42. The table value is 6.94. The Between-groups df = 2, and the Residual (Error) df = 4. We fail to reject the null hypothesis. The F ratio (2.42) is not equal to or greater than the table value (6.94). The three methods are not significantly different.

Problem 3. A tennis director wants to see if there is a difference in the lasting quality of tennis balls made by three different companies. Twelve tennis players are matched on their ability to play tennis. They are arranged in blocks and assigned at random one subject to each tennis manufacturing company. The tennis players play with the same ball and the number of sets played before the ball goes dead is recorded.

Tennis Sets

Block	Treatment Condition		
	1	2	3
1	20	50	80
2	25	60	100
3	30	80	85

Level of significance is .01.

Null hypothesis states there is no significant difference between the three tennis manufacturing companies.

Alternative hypothesis states that the three companies are different.

Answer. F = 33.82. The table value is 18.00. The Between-groups df = 2, and the residual (error) df = 4. We reject the null hypothesis, because the F ratio (33.82) is greater than the table value (18.00). The three tennis balls from different manufacturers are significantly different. The Randomized Blocks Design, like the Randomized Groups Design, does not tell you which pair of groups are significantly different. We again must use a significance test like the Scheffé test.

TEST 7

Directions: Fill in the blanks with the names of the appropriate tests. Read each description carefully.

1. Interval level of measurement, five related groups: _____.
2. Ordinal level of measurement, five related groups: _____.
3. Interval level of measurement, five independent groups: _____.
5. Interval level of measurement, two independent groups: _____
6. Interval level of measurement, one group: _____.

ANSWERS

1. Randomized Blocks Design
2. Friedman test
3. Randomized Groups
4. Kruskal-Wallis test
5. t test(II)
6. t test (I)

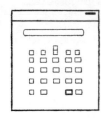

Chapter 8
Correlational Techniques

When we wish to find out the degree to which two variables are related,* we find a *correlation coefficient*. The correlation coefficient gives evidence whether or not a relationship does exist between the variables. We discuss correlation when we talk usually of strength and sign of a relationship. We never talk about one variable *causing* the other. We cannot infer causality.

We choose correlational techniques on the basis of the situation. Such factors as number of variables, the level of measurement, the kind of techniques (parametric or nonparametric) are examined. This chapter contains four widely used correlational techniques: the *contingency coefficient*, the *Spearman Rank Coefficient of Correlation*, the *Pearson Product-Moment Correlation* and a *simple multiple correlation*.

The contingency coefficient is a nonparametric technique which measures the extent of association between two variables when we have nominal level of measurement. The Spearman Rank is a nonparametric measure which measures the extent of association between two variables when both variables are measured at least at the ordinal level of measurement. The Pearson Product-Moment Correlation is a parametric measure where the two variables are linear and are measured at least at the interval level of measurement.

The multiple correlation is a parametric technique which uses the Pearson. This technique involves more than two variables. It is discussed here in its simplest form that is the relationship between one variable and a combination of two other variables.

*Correlation is basically a measure of the relationship between two variables.

CONTINGENCY COEFFICIENT

The contingency coefficient is a measure of correlation that tells the extent of a relationship between two sets of variables. It is a nonparametric measure that involves only two variables. The data is nominal and categorical. This coefficient ranges from 0 and it never quite reaches +1. It is never negative. The sample size must have no more than 20% of the expected frequencies smaller than 5. Each person in the study is classified in one category on each variable. We do not refer to the sampling distribution for the contingency coefficient, instead we refer to the chi-square distribution. We also use the chi-square in computing the contingency coefficient.

☐ **Formula: Contingency Coefficient.**

$$C = \sqrt{\frac{\chi^2}{\chi^2 + N}}$$

Example. A researcher wanted to find out if there was a relationship between college curriculum choice and political party preference. He choses a random sample of 74 students and classified them on the basis of their curriculum choice and their political party preference. The data are recorded in the table below.

Level of significance is .05.

Null hypothesis states there is no relationship between the two variables.

Alternative hypothesis states that political party and curriculum choice are related.

	Republican	Democrat
College	20	6
General	9	14
Vocational	5	20

Procedure	Record	Press	Display
1. Find N. Record the total number of subjects in the study. Write opposite N =.	N = 74	20 + 9 + 5 + 6 + 14 + 20 =	74
2. Find χ^2. Refer to			

Procedure	Record	Press	Display
Chapter 3 for the procedure for χ^2 (II). Record the value you then find after you compute χ^2 for the above problem. Write the answer opposite χ^2 =____ in the record column, and put the value in memory.	$\chi^2 = 16.45$	16.45 M+	16.45
3. Find $\chi^2 + N$. Add N to χ^2 and write the answer in the record column opposite $\chi^2 + N =$ ____.	$\chi^2 + N = 90.45$	RM/CM + 74 =	90.45
4. Divide χ^2 by $\chi^2 + N$ (refer to the record column opposite 3); take the square root of this value. Contingency Coefficient The display is the con-		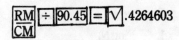	.4264603

| Procedure | Record | Press | Display |

tingency coefficient. Write C = .4264603 in the record column opposite C = _____. Round it to its nearest hundredths place.   C = .43

Statistical Decision. Write the answers in the blank to the left.

$\chi^2 = \underline{16.45}$ 1. Record the chi square value. Round it to the nearest hundredths.

df = $\underline{2}$ 2. Find df. Use the formula (rows − 1) × (columns − 1). Subtract 1 from the number of columns (2 − 1) and subtract 1 from the number of rows (3 − 1), then multiply the differences together. Write the answer opposite df = ____ in the blank to the left.

Table value = $\underline{5.99}$ 3. Find the Table Value by entering Table A with your df(Step 2). Run your fingers across this row until you are under the predetermined level of significance (.05). Write 5.99 in the blank.

reject 4. If your chi-square value (16.45) is equal to or greater than the table value (5.99), reject the null hypothesis. Write reject.

We reject the null hypothesis which says that the relationship between the two variables was 0. There appears to be a relationship between curriculum choice and political party preference.

The C value of .43 gives you the magnitude of the association between the variables.*

Problem 1. A high-school teacher from St. Louis wants to see if there is a correlation between sex and passing or failing a chemistry test. She chose a random sample of 118 students and

---

*For a complete interpretation of C refer to *Statistics for the Social Sciences* by Vicki F. Sharp, Little, Brown & Company, 1979, pages 334-336.

classified them as male or female and as pass or fail on the chemistry entrance test. The data are recorded in the table below.

Level of significance is .01.

Null hypothesis states that there is no relationship between two variables.

Alternative hypothesis states that there is a relationship between gender and passing or failing the chemistry test.

Sex	Pass	Fail
Female	60	10
Male	33	15

Answer. We fail to reject the null hypothesis. $\chi^2 = 5.28$. The table value is 6.64. $C = .21$. There appears to be no relationship between sex and passing the chemistry test.

Problem 2. A psychiatric counselor wants to see if there is a correlation between sex and curriculum choice. She chose a random sample of 147 students and classified them as male or female and class preference as art, shop, or home economics. The data are recorded in the table below.

Level of significance is .05.

Null hypothesis states that there is no relationship between the two variables.

Alternative hypothesis states that there is a relationship between gender and class preference.

	Female	Male
Art	20	20
Shop	30	30
Home Economics	22	25

Answer. We reject the null hypothesis. $\chi^2 = 12.37$. The table value is 5.99. $C = .28$. There appears to be a relationship between gender and class preference.

Problem 3. A college instructor wants to see if there is a correlation between gender and cramming for an exam. He randomly selected 142 students. He then recorded whether a student crammed for his last exam. The data are recorded in the table below.

Level of significance is .01.

Null hypothesis states there is no correlation between cramming for an exam and gender.

Alternative hypothesis states that there is a relationship between cramming for an exam and gender.

	Male	Female
Cramming	10	52
Studying Ahead	60	20

Answer. We reject the null hypothesis. $\chi^2 = 37.90$. The table value is 6.64. C = .34 There appears to be a relationship between cramming for an exam and gender.

SPEARMAN RANK COEFFICIENT (rho)

The Spearman rank coefficient* is a nonparametric measure for use with data that is in the form of ranks or is reduced to ranks. It is used with ordinal data. There are two variables, and each person in the study is ranked separately on each variable. This coefficient ranges from +1 to −1.

☐ **Formula: Spearman Rank Coefficient.**

$$\rho = 1 - \frac{6\Sigma D^2}{N(N^2 - 1)}$$

Example. Tom Brown wanted to find out if there was a correlation between beauty and ballet ability. After a talent contest, he collected the judges scores on appearance and dancing ability. Each girl's score is recorded in the table below.

Level of significance is .10.

Null hypothesis states that there is no relationship between the two variables; that there is no correlation between beauty and ballet ability.

Alternative hypothesis states that beauty and ballet ability are related. A two-tail test is used.

Subject	Beauty	Rank	Ballet Ability	Rank
A	35	4	23	1
B	25	2	40	3
C	30	3	35	2
D	20	1	50	4

Instructions: Note the variables are ranked separately.

---

*If you require a formula that corrects for ties, use $r = \dfrac{\Sigma x^2 + \Sigma y^2 - \Sigma d^2}{2\sqrt{\Sigma x^2 \Sigma y^2}}$

as explained in *Nonparametric Statistics* by Sidney Siegel, McGraw Hill, 1956, pages 206-210.

Procedure	Record	Press	Display
1. Find N. Count the number of people in the group. **Write** the answer in the record column opposite N = ___.	N = 4		
2. Find ΣD^2. Subtract the ranks for each person, square the difference, then put the results in memory to be added.		$\boxed{4}\boxed{-}\boxed{1}\boxed{\times}\boxed{=}\boxed{M^+}$	9
		$\boxed{2}\boxed{-}\boxed{3}\boxed{\times}\boxed{=}\boxed{M^+}$	1
		$\boxed{3}\boxed{-}\boxed{2}\boxed{\times}\boxed{=}\boxed{M^+}$	1
		$\boxed{1}\boxed{-}\boxed{4}\boxed{\times}\boxed{=}\boxed{M^+}$	9
3. Find $6\Sigma D^2$. *a.* Recall from memory the ΣD^2 and clear just memory.		$\boxed{\text{RM/CM}}\boxed{\text{RM/CM}}$	20
b. Multiply the display by 6 and put the answer in memory.		$\boxed{\times}\boxed{6}\boxed{=}\boxed{M^+}$	120
4. Find $N(N^2 - 1)$. *a.* Square N (refer to the record column opposite 1) then subtract 1, then multi-		$\boxed{4}\boxed{\times}\boxed{=}\boxed{-}\boxed{1}\boxed{\times}\boxed{4}\boxed{=}$	60

	Procedure	Record	Press	Display
	ply the results by N. *b.* Write this answer in the record column opposite $N(N^2-1)$ ___	$N(N^2-1) = 60$		
5.	Find $\dfrac{6\Sigma D^2}{N(N^2-1)}$.			
	a. Recall from memory $6\Sigma D^2$ and clear just memory.		RM RM / CM CM	120
	b. Divide the number in display by $N(N^2-1)$ (refer to 4b) and store this in memory.		÷ 60 M⁺	2
6.	Find ρ.			
	a. Subtract the memory from one.		1 − RM/CM =	−1.0
	b. Write the number in display as the Spearman Rank Coefficient. Write $\rho = -1.0$ in the record column opposite =.	$\rho = -1.0$		

Statistical Decision. Record your answers in the blank to the left.

Instructions: When your sample sizes are from 4 to 10 use Table K, if not use Table L.

$\rho = \underline{-1.0}$ 1. Record your correlation in the blank to the left opposite $\rho =$ ___.

$N = \underline{4}$ 2. Record the value you found for N in the blank to the left opposite $N =$ ___.

Table Value = $\underline{1.000}$ 3. Enter Table K. Run your fingers down the N column until you arrive at the value you found for N(4); next, go across this row until you are under your predetermined level of significance (.10) and proper tail (two-tail). Write 1.000 as your table value in the blank to the left.

<u>Reject</u>   4. If the ρ value you found (−1.00) is equal to or more than the table value, reject the null hypothesis. Write reject in the blank to the left.

Since it is a perfect negative correlation, the null hypothesis is rejected. The alternative hypothesis seems to be true—beauty and ballet ability seem to be related.

Problem 1. Mr. Rodriguez, a sociologist, wanted to see if there was a relationship between a person's score on a psychological inventory for the assessment of attitudes and their talent score as assessed by judges. Mr. Rodriguez recorded the scores of 11 people who took both tests as shown in the accompanying table.

Student	Attitude	R_1	Talent	R_2
1	20	4	11	3.5
2	10	2	8	2
3	4	1	40	11
4	33	7.5	1	1
5	24	6	33	9
6	66	11	22	7
7	43	10	12	5.5
8	22	5	11	3.5
9	40	9	36	10
10	33	7.5	24	8
11	15	3	12	5.5

Level of significance is .01.
Null hypothesis states there is no correlation between the two variables.
Alternative hypothesis states there is a positive relationship between the two variables. A one-tail test is used.

Answer. $\rho = .10$. You see Table L because there are 11 subjects in your sample. You use a df $= N - 2$. If ρ is equal to or greater than your table value, reject the null hypothesis. Your df is 9 and your table value is .6851. Since .10 is not equal to or greater than .6851, we fail to reject the null hypothesis. There appears to be no relationship between talent and attitude.

Problem 2. Ms. Angelo, a flying instructor, wanted to see if there was a relationship between the scores earned on a simulated flying test and the scores earned on a real flying test. The simulated flying test consisted of a control wheel, pedals, panel, and a video tape showing different flying situations. The person in the simulator reacted by turning the wheel and working the pedals. The actual flying test had the person flying different maneuvers. Ms. Angelo recorded the scores of eight people who took both tests as shown in the accompanying table.

Individual	Mock Test	R_1	Actual Test	R_2
A	50	4	56	5
B	52	5	58	6
C	60	6	70	7
D	61	7	52	4
E	40	3	50	3
F	35	2	37	1
G	29	1	40	2
H	90	8	95	8

Level of significance is .05.

Null hypothesis states there is no correlation between the two variables.

Alternative hypothesis states there is a positive relationship between the scores on the two tests. A one-tail test is used.

Answer. $\rho = .83$. You use Table K, because there are eight subjects in your sample. The table value is .643. You reject the null hypothesis, since the value for $\rho(.83)$ is greater than the table value (.643). There appears to be a positive relationship between the two variables.

Problem 3. A panel of computer executives wanted to see if there was a relationship between the quality of advertising and the quality of their computer. They randomly selected 7 computers for their study. Each computer was ranked from 1 to 25 on its quality and also ranked from 1 to 25 on the quality of the advertisement used to promote the computer. The rankings are shown in the following table.

Computer	Product	R_1	Advertisement	R_2
A	25	7	10	4
B	22	5	8	3
C	11	3	24	7
D	6	2	20	6
E	24	6	3	1
F	15	4	12	5
G	4	1	5	2

Level of significance is .02.
Null hypothesis states there is no correlation between the scores on the two variables.
Alternative hypothesis states there is a relationship between the two variables.

Answer. $\rho = -.29$. You use Table K, because there are seven subjects in your sample. The table value is .893. You fail to reject the null hypothesis, since the value for ρ ($-.29$) is not equal to or greater than the table value (.893). There appears to be no relationship between the two variables.

When N is 30 or more. If N is more than 30 after you find ρ you have to use another formula, which is $z = \rho\sqrt{N-1}$.

To find the table value, you go to Table D. You find the table value by first finding your z value that you computed in the z column. For example, let's say you have a z value of 3.0, a level of significance of .05, and a two-tail test. You run your fingers down the z column till you come to 3.0, then you go across this row and under .00 you find a probability of .0013. For a two-tail test, you must double this probability. Therefore, .0013 becomes .0026. If this probability (.0026) is equal to or less than your level of significance (0.5) you reject the null hypothesis. In this case you reject.

PEARSON PRODUCT-MOMENT CORRELATION

The Pearson Product-Moment Coefficient of correlation measures the linear relationship between two numerical random variables. When we assume *linearity* we mean that the values for X and Y tend to scatter along a straight line. Figure 8-1 illustrates this.

The Pearson Product-Moment Coefficient of correlations is a parametric measure which assumes the scores for each variable is from a normal distribution. This coefficient ranges from a -1 to a $+1$. The shape of both score distributions should be similar and

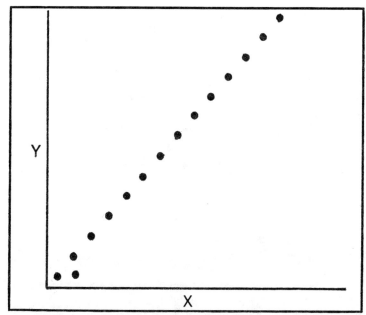

Fig. 8-1. A linear relationship between two numerical random variables.

they cannot be skewed distributions. The scores are at the interval level of measurement. Another condition that should be met in order to use the Pearson is *homoscedasticity*. In order for you to determine if this condition is met, you could draw a scattergram and inspect it. The distance of the dots on this scattergram should be nearly the same on both sides of the line. Figure 8-2 shows a homoscedastic relationship.*

☐ Formula: Pearson Product-Moment Coefficient

$$r = \frac{N\Sigma XY - (\Sigma X)(\Sigma Y)}{\sqrt{[N\Sigma X^2 - (\Sigma X)^2][N\Sigma Y^2 - (\Sigma Y)^2]}}$$

Example. A psychometrist wanted to test for a relationship between math and music. She randomly selected five individuals and measured them on these two variables. She recorded the test scores for each individual in the table below.

---
*If you want to further test for linearity or find more sophisticated methods for determining homoscedastic relationships consult an advanced text such as Lindgren, B.M., *Statistical Theory*, 3rd ed. New York: Macmillan, 1976.

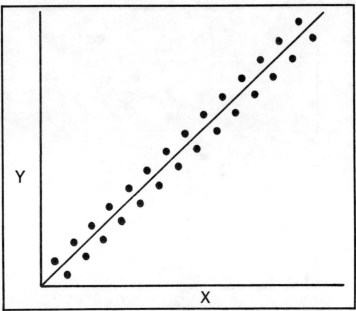

Fig. 8-2. A scattergram helps to determine the condition homoscedasticity.

Level of significance is .05.
Null hypothesis states there is no relationship between the two variables.
Alternative hypothesis states that math and music are related. A two-tail test is used.

Math and Music Test Scores

Students	Math X	Music Y
A	4	9
B	2	3
C	6	7
D	2	9
E	2	9

Instructions: You will put your answers in Box A, Box B or Box C, then in the summary table. Later you will transfer the answers to the r formula.

Procedure	Record	Press	Display

$\boxed{A}$ = NΣXY

	Procedure	*Record*	*Press*	*Display*
1.	Find N. Count the number of individuals in the study. Write 5 opposite N=___.	N = 5		
2.	Find ΣXY. Multiply each X score by its corresponding Y score and put the answer in memory plus to be added.		[4]×[9][M+] [2]×[3][M+] [6]×[7][M+] [2]×[9][M+] [2]×[9][M+]	36 6 42 18 18
3.	Find NΣXY.			
a.	Multiply N (refer to A.1) times memory and put the answer in Box A.		[5]×[RM/CM][=]	600
b.	Clear memory and display.		[RM/CM][RM/CM][C]	0

$$\boxed{A = 600} = N\Sigma XY$$

Now transfer the answer (600) to the summary table under Box A.

	Procedure	*Record*	*Press*	*Display*

$\boxed{B} = (\Sigma)(\Sigma X)$

1. Find ΣX.
   a. Add all the numbers the X column, and put the sum in memory plus.

[4]+[2]+[6]+[2]+[2][M+]   16

Procedure	Record	Press	Display
b. Record the sum also opposite $\Sigma X = $ ___.	$\Sigma X = 16$		
2. Find ΣY.			
a. Add all the numbers in the Y column.		9 + 3 + 7 + 9 + 9 =	37
b. Write the answer opposite $\Sigma Y = $ ___.	$\Sigma y = 37$		
c. Multiply the ΣY by memory and write the results in Box B.		× RM/CM =	592

$\boxed{B = 592}$ $(\Sigma X)(\Sigma Y)$

Now transfer the answer (592) to the Summary Table under Box B.

Procedure	Record	Press	Display
$\boxed{C} = [N\Sigma X^2 - (\Sigma X)^2]$			
1. Find $(\Sigma X)^2$.			
a. Square the value in memory write it opposite $(\Sigma X)^2 = $ ___.	$(\Sigma X)^2 = 256$	RM/CM × =	256
b. Clear the memory and display.		RM/CM RM/CM C	0
Find ΣX^2.			2.
a. Square each score in the X column		4 × = M+ 2 × = M+ 6 × = M+	16 4 36

Procedure	Record	Press	Display
and put each one in memory plus to be added.		`2 × = M+` `2 × = M+`	4 4
b. Multiply N (refer to A.1) by memory, then subtract $(\Sigma X)^2$ (refer to C.1) from the answer.		`5 × RM/CM − 256 =`	64
c. Write the answer in Box C. Clear memory and display.		`RM/CM` `RM/CM` `C`	0

$$\boxed{C = 64} = [N\Sigma X^2 - (\Sigma X)^2]$$

Now transfer the answer (64) to the summary table under Box C.

Procedure	Record	Press	Display
$\boxed{D} = [N\Sigma Y^2 - (\Sigma Y)^2]$			
1. Find $(\Sigma Y)^2$. Find ΣY (refer to B.2) and square it. Write the answer opposite $(\Sigma Y)^2 = \underline{\quad}$.	$(\Sigma Y)^2 = 1369$	`37 × =`	1369
2. Find ΣY^2.			

a. Square each score in the Y column and put the answer in memory plus memory to be added.		`9 × = M+`
		`3 × = M+`
		`7 × = M+`
		`9 × = M+`
		`9 × = M+`
			81
			9
			49
			81
			81

Procedure	Record	Press	Display
b. Multiply N (refer to A.1) by memory, then subtract the $(\Sigma Y)^2$ (refer to D.1) from the answer.		5 × RM/CM − 1369 =	136
c. Write the answer in Box D, then clear memory and display.		RM/CM RM/CM C	0

$$\boxed{D = 136} = [N\Sigma Y^2 - (\Sigma Y)^2]$$

Now transfer the answer (136) to the summary table under Box D.

Instructions: We will now substitute in the formula

$$r = \frac{N\Sigma XY - (\Sigma X)(\Sigma Y)}{[N\Sigma X^2 - (\Sigma X)^2][N\Sigma Y^2 - (\Sigma Y)^2]}$$

from the summary table.

Summary Table

A	B	C	D
600	592	64	136

$$r = \frac{A - B}{\sqrt{(C)(D)}} = \frac{600 - 592}{\sqrt{(64)(136)}}$$

Procedure	Record	Press	Display
1. Subtract B from A and write the answer opposite $N\Sigma XY - (\Sigma X)(\Sigma Y)$ in the record column.		600 − 592 =	8
2. Multiply C by D, then	$N\Sigma XY - (\Sigma X)(\Sigma Y) = 8$	64 × 136 = √ M+	93.29523

Procedure	Record	Press	Display

find the square root of your answer, put this answer in memory.

3. Divide NΣXY − (ΣX)(ΣY) (record column) by memory plus.

Press: `8 ÷ RM =` / `CM`   Display: .0857492

4. The Pearson r = .0857492.

Statistical Decision. Record the answers in the blank to the left.

r = <u>.09</u>   1. Record the value found for r in the blank to the left. Round it to the nearest hundredth place.

df = <u>3</u>   2. Find df. Use the formula N − 2. Subtract 2 from 5 and write 3 opposite df = ___.

Table value = <u>.8783</u>   3. Find your table value. Enter Table L with your df of 3. Run your fingers across the row until you are under the predetermined level of significance (.05) and the proper tail (two-tail). Write the table value .8783 in the blank to the left.

<u>Fail to reject</u>   4. If your r value (absolute value) is equal to or greater than the table value, reject the null hypothesis. Write fail to reject in the blank to the left. There is no relationship between math and music ability.

Problem 1. A professor wanted to find out if there was a correlation between performance in English and performance in biology. He selected at random four students and recorded their test scores for biology and English. The data is shown in the following table:

Student	Biology X	English Y
A	5	8
B	10	20
C	25	15
D	20	13

Level of significance: .01.
Null hypothesis states there is no correlation between the two variables.
Alternative hypothesis states there is a positive relationship between the two variables. A one-tail test is used.

Answer. r = .26. You use Table L. The degrees of freedom equal 2. Your table value is .9800. Since .26 is not equal to or greater than .98, we fail to reject the null hypothesis. There appears to be no relationship between ability in biology and ability in English.

Problem 2. Chancey and Ali wanted to find out if there is a correlation between childrens' weight and strength. They randomly selected six children and measured them on these two variables. They recorded their scores on both tests. The results are shown in the following table.

Child	Weight	Strength
A	100	60
B	70	55
C	65	40
D	60	30
E	55	30
F	66	50

The *Level of significance* is set at .01.
The *Null hypothesis* states there is no correlation between the two variables.
The *Alternative hypothesis* states there is a relationship between the two variables. A two-tail test is used.

Answer. r = .81. You use Table L. The degrees of freedom equal 4. Your table value is .9172. Since .81 is not equal to or greater than .9172, we fail to reject the null hypothesis. There appears to be no relationship between weight and strength.

Problem 3. Ruth and Mary want to test the hypothesis that the longer a person takes to complete a statistics test, the lower their score. They randomly selected eight students and recorded the number of hours it took to complete a statistics test and the grade the student received. The results are shown in the following table.
The *Level of significance* is set at .05.
Null hypothesis states there is no correlation between the two variables.

Grade for Statistics Test and Number of Hours to Complete Test

Student	Score X	Hours Y
A	70	4
B	60	4
C	90	3
D	95	2
E	95	1
F	95	1

Alternative hypothesis states there is a negative relationship between the two variables. A one-tail test is used.

Answer. $r = -.88$. You use Table L. The degrees of freedom equal four. Your table value is .7293. Since $-.88$ is greater than .7293, we reject the null hypothesis. There appears to be a relationship between the two variables.

When the Sample (N) is 30 or More. If N is more than 30 after you find r, you have to use another formula which is
$$z = r\sqrt{N-1}$$

To find the table value, you go to Table E. You find the table value by first finding the z value that you computed in the z column. For example, let's say you have a z value of 2.0, a level of significance of .05, and a two-tail test. You run your fingers down the z column till you come to 2.0, you then go across this row and under .00 you find a probability of .0228. For a two-tail test, you must double this probability. Therefore, .0228 becomes a .0456. If this probability (.0456) is equal to or less than your level of significance (.05) you reject the null hypothesis. In this case you reject.

SIMPLE MULTIPLE CORRELATION

Multiple correlation is a parametric measure involving three variables. In its simplest form it is used when we wish to know the relationship between one variable and two other variables taken together. In a typical situation involving multiple correlation, the first variable or set of scores might be freshman grade point average. This first variable is called the *criterion variable*. The other two variables might be high school grades and college entrance exam scores, both called *predictor variables*.

☐ Formula: Multiple Correlation.

$$R_{1.23} = \sqrt{\frac{r_{12}^2 + r_{13}^2 - 2r_{12}r_{13}r_{23}}{1 - r_{23}^2}}$$

Example. A college professor wants to see if there is a relationship between the criterion variable (variable one) freshman college grade point average and two predictor variables (variable two) high school grades and (variable three) college entrance exam scores. She performs a Pearson Product-Moment Correlation first between variable 1 and variable 2 (r_{12}) (freshman college GPA and high school grades) and then she performs a Pearson between variable 1 and variable 3 (r_{13}) (freshman college GPA and college entrance exam scores), and then she performs a Pearson between variable 2 and variable 3 (r_{23}) (high school grades and college entrance exam scores). For this problem we will assume the correlations were:

$$r_{12} = +.70 \quad R_{13} = +.40 \text{ and } r_{23} = +.80$$

Instructions: The correlations for r_{12}, r_{13} and r_{23} are given. Compute a multiple correlation.

Procedure	Record	Press	Display
1. Find $r_{12}^2 + r_{13}^2$.			
a. Square the value for r_{12} and put it in memory plus to be added.		.70 × = M+	.49
b. Square the value for r_{13} and put it in memory plus to be added.		.40 × = M+	.16
2. Find $2r_{12}r_{13}r_{23}$. Multiply 2 times		2 × .70 × .40 × .80 =	.448

Procedure	Record	Press	Display

the value for r_{12} times this by r_{13}, then times it by r_{23}. Record the answer opposite $2r_{12}r_{13}r_{23}$ = ___. $2r_{12}r_{13}r_{23}$ = .448

3. Find $r^2_{12} + r^2_{13} - 2r_{12}r_{13}r_{23}$. Recall $r^2_{12} + r^2_{13}$ from memory and clear memory, then subtract $2r_{12}r_{13}r_{23}$ (step 2) put the answer in memory plus. | RM/CM | RM/CM | − | .448 | M+ | .202

4. Find r^2_{23}. Square the value for r_{23} and put it opposite r^2_{23} = ___. r^2_{23} = .64 | .80 | × | = | .64

5. Find $1 - r^2_{23}$. Subtract the value for r^2_{23} (step 4) from 1 and write the answer opposite $1 - r^2_{23}$ = ___. $1 - r^2_{23}$ = .36 | 1 | − | .64 | = | .36

Procedure	Record	Press	Display
6. Find $\sqrt{\dfrac{r_{12}^2 + r_{13}^2 - 2r_{12}r_{13}r_{23}}{1 - r_{23}^2}}$			
a. Recall from memory $r_{12}^2 + r_{13}^2 - 2r_{12}r_{13}r_{23}$ and divide it by $1 - r_{23}^2$ (step 5).		RM/CM ÷ .36 =	.5611111
b. Find the square root of the number in display. Record this as $R_{1.23}$ multiple correlation rounding it to the hundredths place $r_{1.23} = .75$.		√ =	.7490734

The multiple correlation between the criterion variable freshman college grade point average and the two predictor variables is .75. You will notice the correlation between freshman GPA and high school grades was +.70 and the correlation between freshman GPA and college entrance exam scores was +.40. When we combine the high school grades with the entrance exam scores, the multiple correlation produces a higher correlation than the simple correlation between freshmen GPA and college entrance exam scores and the correlation between GPA and high school grades.

Problem 1. An admissions officer wants to see if there is a relationship between the criterion variable of freshman grades, variable one and the combined effects of the scores on the Ohio State psychological exam, variable two and the Scholastic Aptitude Test, variable three. He performs simple correlations between the variables. The results follow:

$$r_{12} = .52 \quad r_{13} = .58 \quad r_{23} = .40.$$

Using the preceding information compute a multiple correlation.

Answer. $r_{1.23} = .66$.

Problem 2. A researcher wants to see if there is a relationship between the criterion variable high school grade point average, variable one and the combined effects of the numerical ability portion of the Differential Aptitude Test, variable two and the Iowa tests of Educational Development, variable three. He performs simple correlations between the variables. The results follow:

$$r_{12} = .60 \quad r_{13} = .45 \quad r_{23} = .40.$$

Using the preceding information compute a multiple correlation.

Answer. $r_{1.23} = .64$.

Problem 3. A medical school officer wants to see if there is a relationship between the criterion variable freshman medical school GPA, variable one, and the combined effects of the Wechsler-Bellevue Scale Scores, variable two and the College Entrance Examination Test, variable three. He performs simple correlation between the variables. The results follow:

$$r_{12} = .80 \quad r_{13} = .55 \quad r_{23} = .35.$$

Using the preceding information, compute a multiple correlation.

Answer. $r_{1.23} = .85$.

TEST 8

Directions: Fill in the blanks with the appropriate response.
1. Linear
   Scores from normal distribution
   Level of measurement is interval
   Scattergram is homoscedastic _____
2. Ordinal data
   Nonparametric measure
   Two variables _____
3. Nominal data
   Two variables
   Nonparametric measure _____
4. The contingency coefficient uses the _____ distribution.

5. The _____ correlation technique is a parametric technique.
6. Interval data
   Three variables
   Parametric measure _____

ANSWERS
1. Pearson Product Moment
2. Spearman Rank Coefficient
3. Contingency Coefficient
4. Chi Square
5. Pearson Product-Moment
6. Multiple Correlation

Chapter 9
Other Correlational Techniques

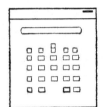

In Chapter 8 there was a discussion of the most popular correlation techniques. However, there are times when other correlational techniques are more applicable to special situations. For instance, when one variable is continuous and the other is dichotomous we use a technique called the *Point Biserial*. It is a special case of the Pearson. When you need a nonparametric measure which measures the relationship between three or more variables, you use the *Kendall Concordance*. When you need a nonparametric test that can be generalized to a partial correlation coefficient, you would use the *Kendall Rank Correlation Coefficient (tau)* and not the Spearman Rank. In the following chapter we will be discussing these techniques and you will learn how to compute them using the calculator.

POINT-BISERIAL CORRELATION

The point biserial correlation is used quite often in test construction analysis. It is a special case of the Pearson Product Moment Correlation where one variable is continuous and the other is dichotomous. A continuous variable is one that is represented by the normal distribution and measured on the interval or ratio scale. For example, achievement test scores and students' heights expressed in inches are continuous variables. A dichotomous variable represents a real dichotomy on a nominal scale. For example, male-female married-unmarried represent true dichotomies. In order to facilitate computation of the point biserial correlation coefficient, the dichotomous variable is assigned values of 1 and 0.

☐ **Formula: Point-Biserial.**

$$r_{pb} = \frac{\overline{Y_1} - \overline{Y_0}}{\sqrt{\frac{\Sigma Y^2}{N-1} - \frac{(\Sigma Y)^2}{N(N-1)}}} \sqrt{\frac{N_1 N_0}{N(N-1)}}$$

Examples. An instructor wanted to determine the relationship between physics exam scores and gender. Eleven students were chosen at random and their physics exam score and gender were recorded. The continuous variable for these scores is the physics exam labeled Y. The dichotomous variable is gender. The male student is assigned a value of 1 and the female student is assigned a value of 0. The results are recorded below.

Level of significance is .05.

Null hypothesis states that there is no relationship between gender and physics test scores.

Alternative hypothesis states that physics and gender are related. A two-tail test is used.

Student	Exam Scores Y	Gender
A	100	1
B	95	0
C	72	1
D	73	0
E	60	0
F	85	1
G	98	1
H	75	0
I	85	1
J	88	1
K	60	0

Instructions: Using the table above, create a second table. First, make a list of all the physics scores continuous variable that are paired with the gender category 1 and label it Y_1. Second, make another list of all the physics scores that are paired with gender category 0 and label it Y_0.

Y_1 Scores Paired with Gender 1	Y_0 Scores Paired with Gender 0
100	95
72	73

Y_1 Scores Paired with Gender 1	Y_0 Scores Paired with Gender 0
85	60
98	75
85	60
88	

You will put your answers in Box A, Box B, or Box C, then in the summary table. Later, you will transfer the answers to the r_{pb} formula.

Procedure	Record	Press	Display
$\boxed{A} = \overline{Y}_1 - \overline{Y}_0$			
1. Find ΣY_1. Add the scores for Y_1. Write the answer opposite the $\Sigma Y_1 = \underline{}$ in the record column.	$\Sigma Y_1 = 528$	$\boxed{100} + \boxed{72} + \boxed{85} + \boxed{98} +$ $\boxed{85} + \boxed{88} =$	528
2. Find N_1. Count the number of scores in Group Y_1. Write the answer opposite $N_1 = \underline{}$ in the record column.	$N_1 = 6$		
3. Find $\overline{Y}_1$. Divide ΣY_1 (display) by N_1 (record column) then put the answer in memory.		$\boxed{\div} \boxed{6} \boxed{M^+}$	88

Procedure	Record	Press	Display
4. Find ΣY_0. Add the scores for Y_0. Write the answer opposite the $\Sigma Y_0 = \underline{}$ in the record column.	$\Sigma Y_0 = 363$	$\boxed{95}\boxed{+}\boxed{73}\boxed{+}\boxed{60}\boxed{+}\boxed{75}\boxed{+}$ $\boxed{60}\boxed{=}$	363
5. Find N_0. Count the number of scores in Group N_0. Write the answer opposite $N_0 = \underline{}$ in the record column.	$N_0 = 5$		
6. Find $\overline{Y}_0$. Divide ΣY_0 (display) by N_0 (record column); write the answer in the record column opposite $\overline{Y}_0 = \underline{}$.	$\overline{Y}_0 = 72.6$	$\boxed{\div}\boxed{5}\boxed{=}$	72.6
7. Find $\overline{Y}_1 - \overline{Y}_0$. Subtract $\overline{Y}_0$ (A.6) from $\overline{Y}_1$ (memory). Write the answer with its sign in Box A and clear memory and display.		$\boxed{\dfrac{RM}{CM}}\boxed{-}\boxed{72.6}\boxed{=}$	15.4
		$\boxed{\dfrac{RM}{CM}}\boxed{\dfrac{RM}{CM}}\boxed{C}$	0

$$\boxed{A = +15.4} = \overline{Y_1} - \overline{Y_0}$$

Now transfer the answer (+15.4) to the summary table under Box A.

Procedure	Record	Press	Display

$$\boxed{B} = \sqrt{\frac{\Sigma Y^2}{N-1} - \frac{(\Sigma Y)^2}{N(N-1)}}$$

1. Find ΣY^2.
   Square all
   the physics
   scores (Y)
   in the first
   table and put
   each one in
   memory to
   be added.

Press	Display
100 × = M+	1000
95 × = M+	9025
72 × = M+	5184
73 × = M+	5329
60 × = M+	3600
85 × = M+	7225
98 × = M+	9604
75 × = M+	5625
85 × = M+	7225
88 × = M+	7744
60 × = M+	3600

2. Find N.
   Count the
   number of
   students in
   the first table.
   Write the
   answer op-
   posite N =
   _____.   N = 11

3. Find N − 1.
   Subtract 1 | 11 − 1 = | 10
   from N (B.2)
   and write
   the answer
   in display
   opposite
   N − 1 = ___. N − 1 = 10

4. Find
   N(N−1). | × 10 = | 110
   Multiply (N−1)

Procedure	Record	Press	Display
(Display) by N (B.2) Record the answer opposite $N(N-1) =$ ___ .	$N(N-1) = 110$		
5. Find $(\Sigma Y)^2$. *a*. Add all the physics exam scores in the table, Y		`100`+`95`+`72`+`73`+`60`+`85`+`98`+`75`+`85`+`88`+`60`= `×`=	891 793881
b. Square the value in display and write the answer opposite $(\Sigma Y)^2 =$ ___ .	$(\Sigma Y)^2 = 793881$		
6. Find $\sqrt{\dfrac{\Sigma Y^2}{N-1} - \dfrac{(\Sigma Y)^2}{N(N-1)}}$ *a*. Recall memory and clear it. Divide the displayed number by $N-1$ (B.3) and put the answer in memory plus.		`RM/CM` `RM/CM` `÷` `10` `M⁺`	7416.1
b. Divide $(\Sigma Y)^2$ (B.5) by $N(N-1)$ (B.4.). Record the answer opposite $\dfrac{(\Sigma Y)^2}{N(N-1)} =$ ___ .	$\dfrac{(\Sigma Y)^2}{N(N-1)} = 7217.1$	`793881` `÷` `110` `=`	7217.1

Procedure	Record	Press	Display
c. Recall and clear memory. Subtract from the display $\frac{(\Sigma Y)^2}{N(N-1)}$ (record column).		$\boxed{\frac{RM}{CM}}\boxed{\frac{RM}{CM}}\boxed{-}\boxed{7217.1}\boxed{=}$	199
d. Find the square root of the display and put the results in memory plus and also in Box B.		$\boxed{\sqrt{}}\boxed{M^+}$	14.106735

$$\boxed{B = 14.106735} = \sqrt{\frac{\Sigma Y^2}{N-1} - \frac{(\Sigma Y)^2}{N(N-1)}}$$

Now transfer 14.106735 to the summary table under Box B.

Procedure	Record	Press	Display
$C = \sqrt{\frac{N_1 N_0}{N(N-1)}}$			
1. Multiply N_1 (A.2) by N_0 (A.5).		$\boxed{6}\boxed{\times}\boxed{5}\boxed{=}$	30
2. Divide by $N(N-1)$ (B.4).		$\boxed{\div}\boxed{110}\boxed{=}$	.2727272
3. Find the square root of the display.		$\boxed{\sqrt{}}$	.5222328
4. Write the answer in display in Box C.			

$$\boxed{C = .5222328} = \sqrt{\frac{N_1 N_0}{N(N-1)}}$$

Now transfer .5222328 to the summary table under Box C.

We will now substitute in the formula

$$r_{pb} = \frac{\overline{Y}_1 - \overline{Y}_0}{\sqrt{\dfrac{\Sigma Y^2}{N-1} - \dfrac{(\Sigma Y)^2}{N(N-1)}}} \sqrt{\frac{N_1 N_0}{N(N-1)}}$$

from the summary table.

Summary Table

A	B	C
+15.4	14.106735	.5222328

$$r_{pb} = \frac{A}{B} C = \frac{15.4}{14.106735} \cdot .5222328$$

Procedure	Record	Press	Display
1. Divide A by B (memory).		15.4 ÷ RM/CM =	1.0916771
2. Multiply the display by C.		× .5222328 =	.5701095
3. $r_{pb} = .57$ rounded to the hundredths place.			
4. Clear memory and display.		RM/CM RM/CM C	0

Statistical Decision. Record your answers in the blank to the left when instructed.

t = <u>2.08</u> 1. Find t. When we use the Point-Biserial formula we test for significance using the formula which follows.

☐ **Formula: Point-Biserial Correlation Testing for Significance.**

$$t = r_{pb} \sqrt{\frac{N-2}{1 - r_{pb}^2}}$$

Working with the values that we just found in the preceding problem (N = 11 and $r_{pb} = .57$) we substitute these numbers in the formula.

$$t = .57 \sqrt{\frac{11-2}{1-(.57)^2}}$$

Use your calculator to solve for t.

Procedure	Record	Press	Display
1a. Find $N-2$. Subtract 2 from N and record the answer opposite $N-2 =$ _____.	$N-2 = 9$	$\boxed{11}\ \boxed{-}\ \boxed{2}\ \boxed{=}$	9
1b. Find r^2_{pb}. Square r_{pb} and put it in memory plus.		$\boxed{.57}\ \boxed{\times}\ \boxed{=}\ \boxed{M^+}$	.3249
1c. Find $1 - r^2_{pb}$. Recall and clear the memory and subtract the display (r^2_{pb}) from 1, then put this result in memory plus.		$\boxed{1}\ \boxed{-}\ \boxed{\genfrac{}{}{0pt}{}{RM}{CM}}\ \boxed{\genfrac{}{}{0pt}{}{RM}{CM}}\ \boxed{M^+}$	.6751
1d. Find t. Divide ($N-2$) step a by $1 - r^2_{pb}$ (memory). Square root this display and multiply it by r_{pb}; round the answer to the hundredths place and record it opposite $t =$ _____ in the		$\boxed{9}\ \boxed{\div}\ \boxed{\genfrac{}{}{0pt}{}{RM}{CM}}\ \boxed{=}$ $\boxed{\sqrt{}}\ \boxed{\times}\ \boxed{.57}\ \boxed{=}$	13.331358 2.0811915

| Procedure | Record | Press | Display |

left hand column opposite 1.

df = <u>9</u> 2. Find the degrees of freedom.
Use the formula N−2 (refer to 1.a). Write 9 in the blank to the left opposite df = _____.

Table value = <u>2.262</u> 3. Find the table value.
Enter Table G (critical values of t). Run your fingers down the df column until you arrive at 9, next go across until you are at your predetermined level of significance (.05) and proper tail (two-tail). Write 2.262 as your table value opposite table value = _____ in the blank to the left.

<u>Fail to reject</u> 4. If the t value you found is equal to or greater than your table value reject the null hypothesis. You write fail to reject.

There appears to be no relationship between physics test scores and gender.

For each of the following problems create a table as in the example problem above.

Problem 1. A marriage counselor wanted to determine the relationship between math achievement scores and marital status. Ten clients were chosen at random and their math test scores and marital status was recorded. The continuous variable, math scores, is labeled Y. The dichotomous variable is marital status. Married clients are assigned a value of 1 and if they are unmarried they are assigned a value of 0. The results are recorded in the following table.

Clients	Math Scores Y	Marital Status
A	95	1
B	85	0
C	90	1
D	60	1
E	50	0
F	45	0
G	79	1
H	88	1
I	89	1
J	76	0

Level of significance is .05.

Null hypothesis states that there is no relationship between marital status and math test scores.

Alternative hypothesis states that math test scores and marital status are related. A two-tail test is used.

Answer. $r_{pb} = .57$. Using Table G with a df of 8, you find table value equal to 2.306. You have a t value of 1.96. You fail to reject the null hypothesis, since the t value (1.96) is not equal to or greater than the table value (2.306). There appears to be no relationship between the two variables.

Problem 2. A researcher wanted to determine the relationship between scores on the driving exam and matriculation from college. Twelve individuals were chosen at random and their driver's exam scores and academic status were recorded. The continuous variable, driver's exam scores, was labeled Y. The dichotomous variable is academic status. If a person graduated from college they were assigned a value of 1 and if they did not graduate they were assigned a value of 0. The results are recorded in the table which follows.

Adult	Drivers' Exam Scores Y	Academic Status
A	100	1
B	90	0
C	95	1
D	80	1
E	50	0
F	75	0
G	80	1
H	70	0
I	60	0
J	55	0
K	85	1
L	65	1

Level of significance is .05.

Null hypothesis states that there is no relationship between driver's exam scores and academic status.

Alternative hypothesis states there is a positive relationship between driver's exam scores and academic status. A one-tail test is used.

Answer. $r_{pb} = .58$. Using Table G with a df of 10 you find a table value equal to 1.812. You have a t value of 2.25. You reject the null hypothesis, since the t value of 2.25 is greater than your table of 1.812. There appears to be a relationship between driver's exam scores and academic status.

Problem 3. A psychologist wants to determine the relationship between Millers Analogy Exam Scores and smoking status. Ten students were chosen at random and their Millers Analogy Exam Scores and smoking status were recorded. The continuous variable is the Millers Analogy exam scores and it is labeled Y. The dichotomous variable is smoking status. A person is assigned a value of 1 if he smokes and a value of 0 if he is a non-smoker. The results are recorded in the table that follows.

Student	Millers Analogy Scores Y	Smoking Status
A	95	1
B	90	0
C	35	0
D	20	1
E	85	1
F	100	0
G	75	0
H	65	1
I	60	0
J	90	1

The *level of significance* is .01.

The *null hypothesis* states there is no relationship between the Millers Analogy Exam Scores and smoking status.

The *alternative hypothesis* states there is a relationship between the Millers Analogy Exam Scores and smoking status. A two-tail test is used.

Answer. r_{pb} = .02. The table value is equal to 3.355; you enter this table with a df of 8. You have a t value of .06. You fail to reject the null hypothesis, since the t value of .06 is not equal to or greater than the table value of 3.355. There appears to be no relationship between the two variables.

KENDALL COEFFICIENT OF CONCORDANCE (W)

The Kendall Coefficient of Concordance (W) is a nonparametric measure for use with data that is in the form of ranks or reduced to ranks. It is used with ordinal data. When we wish to determine the relationship among three or more variables this is the appropriate technique. Perfect agreement is indicated by +1 and 0 indicates lack of agreement.

☐ **Formula: Kendall Coefficient of Concordance.**

$$W = \frac{\Sigma\left(R_j - \frac{(\Sigma R_j)^2}{N}\right)}{\frac{1}{12} k^2 (N^3 - N)}$$

Example Problem. Three personnel officers who work for a toy manufacturing company are asked to interview and rank separately eight job applicants. The three independent rankings that the personnel officers gave are shown in the following table.

Job Applicant

	a	b	c	d	e	f	g	h	
Officer X	8	5	6	2	3	4	7	1	
Officer Y	2	4	3	1	5	7	8	6	
Officer Z	6	3	1	2	4	5	7	8	
Rank Totals (R_j)	16	12	10	5	12	16	22	15	$\Sigma R_j = 108$

Level of significance is .05.

Null hypothesis states there is no relationship among the rankings of the three officers.

Alternative hypothesis states there is agreement in the ranking of the three officers.

Procedure	Record	Press	Display
1. Find ΣR_j. Add the ranks for applicant a and put this sum in memory plus to be added, and also in the record column under the heading Rank Totals.	*Rank Totals* (a) 16 (b) 12 (c) 10 (d) 5 (e) 12 (f) 16 (g) 22 (h) 15	8 + 2 + 6 M+ 5 + 4 + 3 M+ 6 + 3 + 1 M+ 2 + 1 + 2 M+ 3 + 5 + 4 M+ 4 + 7 + 5 M+ 7 + 8 + 7 M+ 1 + 6 + 8 M+	16 12 10 5 12 16 22 15

Procedure	Record	Press	Display
Continue this procedure until all job applicants' ranks are summed and put in memory plus and the record column.			
2. Find N. Count the number of people being ranked and record the number opposite N = ___.	N = 8		
3. Find $\frac{\Sigma R_j}{N}$. Recall memory and clear, then divide the display ΣR_j by N (step 2). Record the answer opposite $\frac{\Sigma R_j}{N}$ = ___.	$\frac{\Sigma R_j}{N} = 13.5$	RM/CM RM/CM ÷ 8 =	13.5
4. Find $\Sigma(R_j - \frac{\Sigma R_j}{N})^2$. Take each person's rank total (record column		16 − 13.5 × = M⁺ 12 − 13.5 × = M⁺ 10 − 13.5 × = M⁺ 5 − 13.5 × = M⁺	6.25 2.25 12.25 72.25

Procedure	Record	Press	Display

step 1), subtract it from $\frac{\Sigma R_j}{N}$ (step 3), square the results, and put it in memory plus to be added.		`12` `−` `13.5` `×` `=` `M+`
		 `16` `−` `13.5` `×` `=` `M+`
		 `22` `−` `13.5` `×` `=` `M+`
		 `15` `−` `13.5` `×` `=` `M+`
			2.25
			 6.25
			 72.25
			 2.25

5. Find $(N^3 - N)$. Take the third power of N (step 2) and subtract N (step 2) from the results. Record the answer opposite $(N^3 - N)$ = ___.	$(N^3 - N) = 504$	`8` `×` `=` `=` `−` `8` `=`	504
6. Find k. Record the number of sets of ranks or number of people doing the ranking opposite k = ___.	k = 3		
7. Find $\frac{1}{12} k^2 (N^3 - N)$. Square k (step 6), then divide the display by 12. Next,		`3` `×` `=` `÷` `12` `×` `504` `=`	378

| Procedure | Record | Press | Display |

multiply this result by $(N^3 - N)$ (step 5. Record the answer opposite

$\frac{1}{12} k^2(N^3 - N) =$ _____ . $\frac{1}{12} k^2 (N^3 - N) = 378$

8. Find W. Recall memory, clearing it and dividing the display $\Sigma(R_j - \frac{\Sigma R_j}{N})$ by $\frac{1}{12} k^2 (N^3 - N)$ (step 7). Record this value opposite W = ___ .   W = .4656084

Press: |RM/CM| |RM/CM| ÷ 378 =   Display: .4656084

Statistical Decision. If your N is larger* than 7 you use the formula: $\chi^2 = k(N-1)W$ to compute a value for chi square whose significance for a df = N − 1 is tested by referring to Table A.

We will now record our answers in the blank to the left and find out if W is significant.

$\chi^2 = \underline{9.78}$ 1. Find χ^2. Using the formula $k(N-1)W$ compute a value for chi square, and record it rounded to the hundredths place in the blank to the left. We find $k(N-1)W$ equal to 3(8−1).4656084. We compute a value for chi square which is 9.78.

---

*When N is less than 8 you refer to the table found in *Nonparametric Statistics* by Sidney Siegel, McGraw-Hill, 1956 page 286.

df = 7   2. Find df. Use the formula N−1. Subtract 1 from N then write the answer opposite df = ____ in the blank to the left.

Table Value = 5.99   3. Find the table value for your df (step 2). Run your fingers across this row until you are under the predetermined level of significance (.05). Write 5.99 in the blank to the left.

Reject   4. If your chi square value (9.78) is equal to or greater than the table value (5.99), reject the null hypothesis. Write reject.

We reject the null hypothesis which says that there is no relationship among the ranking of the three officers. The officers seem to be applying the same ranking standards in this study. This does not imply that the standard applied are correct, just that there is agreement between the three officers.

Problem 1. Four judges are asked to rank the projects of ten children at a school science fair. The four independent rankings that the judges made are shown below.

Child

	a	b	c	d	e	f	g	h	i	j
Judge W	1	4	5	6	7	8	9	2	3	10
Judge X	3	2	6	8	4	7	10	1	5	9
Judge Y	2	1	3	7	6	8	10	4	5	9
Judge Z	1	2	3	4	9	7	8	5	6	10
Rank Totals (R_j)	7	9	17	25	26	30	37	12	19	38

Level of significance is .01.

Null hypothesis states there is no relationship among the rankings of the four judges.

Alternative hypothesis states there is agreement in the ranking of the four judges.

Answer. W = .8318182. Using Table A with a df of 9, you find a table value equal to 21.67. You have a χ^2 equal to 29.95. You reject the null hypothesis, since the chi square value (29.95) is greater than the table value (21.67). There appears to be a relationship among the rankings of the four judges. The judges seem to be applying the same ranking standards for the students' projects.

Problem 2. Ten mothers and their preschool children attended a class on child rearing for a period of six months. A staff of four psychologists worked with these parents. At the end of the course these four psychologists were asked to rank the ten mothers on how well they felt these mothers would rear their children. These rankings are shown below.

Mother

	A	B	C	D	E	F	G	H	I	J
Psychologist W	1	3	4	5	9	8	6	7	2	10
Psychologist X	2	3	8	4	7	6	10	5	9	1
Psychologist Y	5	2	8	7	1	10	3	9	4	6
Psychologist Z	5	10	9	4	8	6	7	2	3	1

Level of significance is .05.
Null hypothesis states there is no relationship among the four psychologists.
Alternative hypothesis states there is agreement in the ranking of the four psychologists.

Answer. W = .2060606. Using Table A with a df of 9 you will find a table value of 16.92. You have a χ^2 value equal to 7.42. You fail to reject the null hypothesis, since the chi square value 7.42 is not equal to or greater than the table value 16.92. There appears to be no relationship among the rankings of the four psychologists. The psychologists seem not to be applying the same overall ranking.

Problem 3. Three judges were asked to judge a talent contest. There were twelve contestants that were ranked by these three independent judges. The results are shown below.

Talent Contestant

	A	B	C	D	E	F	G	H	I	J	K	L
Judge X	1	12	11	10	9	8	6	7	5	4	2	3
Judge Y	2	1	12	11	10	9	8	7	6	5	3	4
Judge Z	1	12	11	10	9	8	7	6	5	4	3	2

Level of significance is .05.
Null hypothesis states there is no relationship among the rankings of the three judges.
Alternative hypothesis states there is agreement in the ranking of the three judges.

Answer. W = .7886557. Using Table A with a df of 11, you find a table value equal to 19.68. You have a χ^2 value equal to 26.03. You reject the null hypothesis since the chi square value (26.03) is greater than the table value (19.68). There appears to be a relationship among the rankings of the three judges. The judges seem to be applying the same ranking on the contestants.

KENDALL RANK CORRELATION (TAU)

The Kendall tau is a nonparametric measure for use with data that is in the form of ranks.* There are two variables and each person is ranked separately on each variable. The Kendall tau is sometimes used in place of the Spearman Rank which was presented in Chapter 8. Like the Spearman, the tau's coefficient varies from +1 to −1. The advantage it has over the Spearman is it can be generalized to a partial correlation.

Formula: Kendall tau.

$$\text{tau} = \frac{P - Q}{\frac{1}{2} N(N-1)}$$

Example. An experimenter wants to compare two judges on the ranking of eleven contestants in an art show. During the show, he has these judges make their ratings. He then collects the ratings and records them in a table.

Contestant	Judge A	Judge B
A	1	4
B	7	1
C	4	7
D	3	6
E	2	3
F	5	5
G	6	2
H	11	8
I	10	9
J	9	11
K	8	10

Level of significance is .05.

Null hypothesis states that there is no relationship between the ranking of the two judges.

Alternative hypothesis states that there is a relationship between the ranking of the judges. A two-tail test is used.

Instructions: In computing tau, we arrange the first column of ranks (Judge A's) from lowest to highest. This table follows with these values arranged this way.

---

*If you require a formula that corrects for ties use $\tau = \dfrac{S}{\sqrt{\frac{1}{2}N(N-1) - T_x} \sqrt{\frac{1}{2}N(N-1) - T_y}}$ as explained in *Nonparametric Statistics* by Sidney Siegel, McGraw Hill, 1956, pages 217-219.

Contestant	Column 1 Judge A	Column 2 Judge B	Column 3 Ranks Higher	Column 4 Ranks Lower
A	1	4	7	3
E	2	3	7	2
D	3	6	5	3
C	4	7	4	3
F	5	5	4	2
G	6	2	4	1
B	7	1	4	0
K	8	10	1	2
J	9	11	0	2
I	10	9	0	1
H	11	8	0	0

You will notice the second column is not in order. We now determine the extent of this lack of order by counting the number of ranks for each individual that are *below him* which are higher and lower than his rank. We form columns 3 and 4 in the table using this procedure. For example, contestant A has 7 ranks higher than 4; that is, ranks 6, 7, 5, 10, 11, 9, 8. Contestant A has 3 ranks lower than 4; that is, ranks 3, 2, and 1. Contestant K has 1 rank higher than 10, that is 11. Notice that contestant K has only 2 ranks lower than 10, that is 9 and 8, because we do *not* include a rank in our counting that has been considered.

Procedure	Record	Press	Display
1. Compute P. Add the scores in Column 3 (Ranks Higher) record the sum opposite $P =$ ___.	$P = 36$	7+7+5+4+4+4+4+1+0+0+0=	36
2. Compute Q. Add the scores in Column 4 (Ranks Lower) record the sum opposite $Q =$ ___.		3+2+3+3+2+1+0+2+2+1+0=	19

Procedure	Record	Press	Display
3. Compute P−Q. Subtract Q (step 2) from P (step 1) and put the answer in memory plus.		$\boxed{36}\boxed{-}\boxed{19}\boxed{M^+}$	17
4. Find N. Count the total number of subjects in the study and record the answer opposite N = ___.	N = 11		
5. Compute $\frac{[N(N-1)]}{2}$. Subtract 1 from N (step 4) multiply it by N and divide the results by 2. Record the answer opposite. $\frac{[N(N-1)]}{2} = $ ___.	$\frac{[N(N-1)]}{2} = 55$	$\boxed{11}\boxed{-}\boxed{1}\boxed{\times}\boxed{11}\boxed{\div}\boxed{2}\boxed{=}$	55
6. Find tau. Divide P − Q (memory plus) by $\frac{[N(N-1)]}{2}$ (step 5). Record the answer in		$\boxed{\frac{RM}{CM}}\boxed{\div}\boxed{55}\boxed{=}$	.3090909

197

Procedure	Record	Press	Display

display
opposite
T = ___ .   T = .3090909

Statistical Decision. If N is larger* than 10 you compute the value for z by using formula 9-6 and refer to Table D.

☐ **Formula: Significance for Kendall tau.**

$$z = \frac{T}{\sqrt{2(2N+5)/[9N(N-1)]}}$$

Record your answer in the blanks to the left.

$z = \underline{1.32}$ 1. Find z. Using the formula above, compute a value for z and record it rounded to the hundredths place in the blank to the left. We find

$$\frac{.3090909}{\sqrt{2(2(11)+5)/[9(11)(11-1)]}}$$

equals 1.3234492 which is recorded opposite z = ___ .

Probability 2. Find z's probability. You enter Table D and find a z
= .1868   of 1.32 has a probability of occurring .0934. For a two-tail test, double the probability shown in the table and record it opposite Probability = ___ .

Fail to   3. If the probability found is equal to or less than your
reject   level of significance, reject the null hypothesis. Since .1868 is not less than .05, we fail to reject the null hypothesis.

There appears to be no agreement in the ranking of the two judges.

Problem 1. A department chairman in English wants to compare two professors on the ranking of eleven students in an essay writing contest. He had these professors' rankings recorded in the following table:

Student	Professor A	Professor B
A	4	7
B	3	6
C	2	8
D	6	9
E	8	5
F	10	4

---

*When N is less than 11, refer to Table P found on page 284 in Sidney Siegel's *Nonparametric Statistics*, McGraw Hill, 1956.

Student	Professor A	Professor B
G	11	10
H	9	3
I	7	1
J	5	2
K	1	11

Level of significance is .01.

Null hypothesis states there is no relationship between the rankings of the two professors.

Alternative hypothesis states that the rankings between the professors is positive. A one-tail test is used.

Note: Remember to rearrange the first column of ranks from lowest to highest.

Answer. T = −.2363636. We find a z equal to −1.01 and its probability of occurring is .1562. We fail to reject the null hypothesis, since the probability .1562 is not equal to or less than .01. There appears to be no agreement in the rankings of the two professors.

Problem 2. A researcher wants to find out if there is a relationship between aggressive behavior and social class striving. Twelve students were administered a personality test which rated them on aggressiveness and a well known scale that measured social class striving. The students ranks are recorded in this table:

Student	Aggressiveness Scale	Social Class Scale
A	8	9
B	7	6
C	6	7
D	5	4
E	9	8
F	10	11
G	4	5
H	11	10
I	3	3
J	12	12
K	1	2
L	2	1

Level of significance is .05.

Null hypothesis states there is no relationship between aggressiveness and social class striving.

Alternative hypothesis states there is a positive relationship between aggressiveness and social class. A one-tail test is used.

Answer. T = .6727272. We find a z equal to 2.88 and its probability of occurring is .0020. We reject the null hypothesis, since the probability .0020 is less than .05. There appears to be a high positive correlation between aggressiveness and social class striving.

Problem 3. The members of the New York City School of Music asked two judges to rank twelve prospective students on their talent. They wanted to determine the degree of agreement between the two judges. The judges' ratings are recorded in this table:

Student	Judge A	Judge B
A	1	2
B	7	1
C	9	8
D	10	3
E	8	7
F	2	4
G	12	9
H	3	5
I	4	10
J	5	12
K	6	11
L	11	6

Level of significance is .05.
Null hypothesis states there is no relationship between the rankings of the two judges.
Alternative hypothesis states that the ranking between the judges are related. A two-tail test used.

Answer. T = .1818181. We find a z equal to .82 which has a probability of occurring a .2061. Since this is a two-tail test, we double this probability which gives us .4122. We fail to reject the null hypothesis, since the probability .4122 is not equal to or less than .05. There appears to be no agreement in the rankings of the two judges.

TEST 9
Directions: Fill in the blanks with the appropriate answers.
1. The _____ is used when one variable is dichotomous and the other is continuous.
2. The _____ is a nonparametric measure used to determine the relationship among three or more variables.

3. The Kendall tau is used sometimes in place of the _____.
4. The _____ is used quite often in test construction analysis.
5. The _____ indicates perfect agreement by +1 and lack of agreement by 0.

ANSWERS

1. Point Biserial
2. Kendall Coefficient of Concordance
3. Spearman Rank
4. Point Biserial
5. Kendall Coefficient of Concordance

Chapter 10
Test Yourself

You have covered considerable material and learned how to work with 25 formulas which represent a wide spectrum of statistical tests and techniques. Upon seeing a problem, you should now be able to choose the appropriate statistical test and, with the aid of an inexpensive calculator, easily solve your statistical problems. Your statistical problems can range from a simple mean to a more complex one-way analysis of variance.

Following are 40 problems. You are to first determine the test or technique you are to use, then use the statistical technique to solve the problem on your calculator. After you have worked a problem check to see if your answer is right by comparing your solution with the one that follows the problem. If you are incorrect, the appropriate chapter is listed by each solution so that you can refer back in the book for a review of the statistical test or technique.

Problem 1. Four plant supervisors, who work for a huge automobile plant, were asked to rank separately nine men that worked on the assembly line. The general manager wanted to examine the degree of relationship among the ratings of the four plant supervisors. The four independent rankings of the supervisors are shown in the table which follows.

Men on the Assembly Line

	a	b	c	d	e	f	g	h	i
Supervisor W	1	3	2	4	6	5	7	9	8
Supervisor X	6	2	1	4	3	7	5	9	8
Supervisor Y	1	3	2	5	4	6	7	9	8
Supervisor Z	1	3	2	4	5	6	7	8	9

202

The *level of significance* is .05.
Null hypothesis states that there is no agreement among the four plant supervisors in the ranking of the men on the assembly line.
Alternative Hypothesis states there is agreement among the supervisors in ranking the nine men on the assembly line.

Answer. You use the Kendall coefficient of concordance. W = .8645833. Using Table A with a df of 8, you have a table value of 15.51. You have a χ^2 value equal to 22.67. You reject the null hypothesis, since the chi square value 22.67 is greater than the table value 15.51. There appears to be agreement in the ranking of the four plant supervisors. (Chapter 9).

Problem 2. A management consultant wanted to see if three popular vending machine companies dispensed the same amount of soft drink in their machines. He drew a sample of twelve machines—four from each company—and recorded the weights of the amount of soft drink dispensed from each machine.

Soft Drink Machine
Measurement in Ounces

Company A	Company B	Company C
6.8	4.0	6.5
6.5	5.0	6.4
5.8	6.0	5.0
6.0	4.5	4.0

The *level of significance* is .05.
The *null hypothesis* states that there is no significant difference between the three vending machine companies.
The *alternative hypothesis* states the three companies are different.

Answer. You use the one-way analysis of variance, randomized group design. F = 2.49. The Between df = 2, the Within df = 9. We fail to reject the null hypothesis. The F value (2.49) is not equal to or greater than your table value (4.26). There is no significant difference between the three companies (Chapter 7).

Problem 3. In a shopping center, a social scientist took a random sample of workers and asked them if they were happy or unhappy with their life's work. He classified them into four categories: Menial, Blue Collar, Office Worker, and Professional, and recorded the results in the table that follows.

	Menial	Blue Collar	Office Worker	Professional
Happy	20	60	50	80
Unhappy	50	40	45	10

The *level of significance* is set at .01.
The *null hypothesis* states that the four groups are the same.
The *alternative hypothesis* states the groups are different.

Answer. We use the Chi Square (II) test. $\chi^2 = 61.03$. The table value is 11.34 using a df equal to 3. We reject the null hypothesis, since the chi square value (61.03) is greater than the table value (11.34). The groups appear to be different in their job satisfaction. (Chapter 5).

Problem 4. A stock market analyst claims there is a positive relationship between the cost of lumber per square foot and the Dow-Jones Stock average. She investigates six random days on the stock market and records her data in the table that follows.

Lumber (cents)	.45	.54	.50	.42	.48	.38
Dow-Jones Average	880	940	930	925	935	920

Level of significance is .05.
Null hypothesis states that there is no relationship between the two variables.
Alternative hypothesis states that lumber per square foot and the Dow-Jones stock average are positively related. A one-tail test is used.

Answer. We use a Pearson Product-Moment Correlation. r = .40. You use Table L. The degrees of freedom equal 4. Your table value is .7293. Since .40 is not equal to or greater than .7293 we fail to reject the null hypothesis. There appears to be no relationship between lumber per square foot and the Dow-Jones stock average. (Chapter 8)

Problem 5. A pipe store owner has developed four different blends of tobacco: Blend A, Blend B, Blend C and Blend D. As an owner, he does not want to overstock blends that are bought less frequently. He wants to determine whether there is a difference in the customers' preference for his blends of tobacco. He randomly selects 100 customers and asks them each to indicate their tobacco blend preference. The results are shown in the following table.

Tobacco Blend Choice

Brand	Frequency
A	20
B	40
C	20
D	20

The *level of significance* is set at .01.

The *null hypothesis* states that the customers' preferences for brands of tobacco are the same.

The *alternative hypothesis* states there is a difference in the customers' preferences for tobacco brands.

Answer. We use the Chi Square (I) test. $\chi^2 = 12$. The table value is 11.34. Since the chi square value (12) is greater than the table value (11.34), we reject the null hypothesis. The customers do not feel the same about the tobacco blends. (Chapter 5)

Problem 6. An owner of a computing firm is thinking of switching to another, less-expensive computer model. The new computer firm claims it has a computer that will outperform any on the market with less costly repairs. One sample of computers is randomly selected from the old firm and one sample of computers is randomly selected from the new firm. They both are tested and their hours of trouble free service are recorded.

Trouble Free Service Scores
Hours

Old Firm	New Firm
90	50
60	45
100	34
50	60
50	35
70	55
40	30
65	55
70	80
80	85
90	65

The *level of significance* is .05.

The *null hypothesis* states there is no significant difference between computers in their service records.

Alternative hypothesis states that the two groups are different. A two-tail test is used.

Answer. The t(II) test is used. t = 1.97. The table value is 2.086 and the df is 20. We fail to reject the null hypothesis, since the t value (1.97) is not equal to or greater than the table value (2.086). There is no significant difference between the two computer companies service records. (Chapter 7).

Problem 7. A doctor wanted to test the effect of a diet on 20 adults. The adults were matched on height and weight and family history. He assigned one member of each matched pair to Group A and the other member to Group B. Group A received a special diet for six months while Group B just counted calories. At the end of six months the weight loss for each person was recorded.

Weight Loss in Pounds
Six Month Period

Pair	Group A	Group B
A	35	42
B	20	25
C	30	25
D	43	40
E	15	20
F	20	25
G	33	30
H	25	27
I	10	15
J	11	14

The *level of significance* is set at .01.
The *null hypothesis* states there is no significant difference between the two groups.
The *alternative hypothesis* states that the two groups are different. A two-tail test is used.

Answer. We use the t(III). t(III) = -1.57. The table value is 2.262. We fail to reject the null hypothesis. The t value (-1.57) is not equal to or greater than the table value (2.262). There is no significant difference between Group A and Group B. (Chapter 7).

Problem 8. A professional art critic asked two judges to rate the watercolor paintings of ten students. He was anxious to see if the judges agreed on what constituted a good watercolor. The judges' scores are recorded in the table that follows.

Judge A	8	7	2	3	5	4	1	6	9	10
Judge B	8	1	5	10	4	7	9	6	3	2

The *level of significance* is set at .05.
The *null hypothesis* states that there is no relationship between the two judges' scores.
The *alternative hypothesis* states there is a positive relationship between the two variables. A one-tail test is used. Use the Spearman Rank Correlation.

Answer. $\rho = -.62$. The table value is .564 (Table K). We fail to reject the null hypothesis, since the relationship is in the wrong direction. There is no evidence here to suggest that the two judges agree on what constitutes good painting. (Chapter 8).

Problem 9. An art director wants to know if there is any difference in the three art technique courses being taught at his institute. He matches artists on their ability to do certain types of drawing exercises. He then assigns a member of each matched triplet to Art I, another to Art II and a third to Art III. At the end of a year the artists are rated on their performance.

Artist Ratings

Matched Triplet	Art I	Art II	Art III
A	20	50	20
B	10	15	25
C	33	38	50
D	38	20	35
E	7	3	1
F	11	25	22
G	44	32	30
H	18	16	14
I	21	40	38

The *level of significance* is set at .01.

The *null hypothesis* states there is no significant difference in the overall ratings of the three groups.

The *alternative hypothesis* states that the ratings of the three groups are different.

Answer. We use the Friedman test. $\chi r^2 = 1.17$. We use Table H and the χr^2 value of .889 comes closest to your χr^2 value of 1.17 without exceeding it. The probability of .889 occurring is .865 and since this probability is not equal to or less than .01 you fail to reject the null hypothesis. The results show there is no difference among the ratings of the three different art courses. (Chapter 6).

Problem 10. An animal trainer is training monkeys to obey certain commands. The trainer is having his associates use two different methods of reinforcement. He would like to determine which method is working better. He randomly selects 17 monkeys. One group of nine monkeys is trained with periodic reinforcement, (I) while the other group of eight monkeys is trained with immediate reinforcement (II). After five weeks the monkeys are rated on their performance. The results follow.

Monkey's Ratings

I	29	30	31	35	20	28	34	10	
II	10	5	8	15	25	32	22	10	16

The *level of significance* is set at .05.

The *null hypothesis* states there is no significant difference between the two groups of monkeys.

The *alternative hypothesis* states there is a difference between the two groups of monkeys. A two-tail test is used.

Answer. The Mann-Whitney U test is used. U = 13. The table value is 15. We reject the null hypothesis since the value for U(13) is *smaller* than the table value (15). There is a difference between the two groups of monkeys. (When you do the Mann-Whitney U be sure to combine the scores of both groups together when you rank.) (Chapter 6).

Problem 11. An office manager was concerned about the stenographers in her pool's efficiency in the office. She devised a new training program to make them more efficient. To test the program, she randomly selects 12 secretaries and enrolls them. She then has some office managers rate each secretary for three days before they enter the program and for three days after they complete the program. The data follows.

Secretaries' Efficiency

Secretary	Before	After
A	72	70
B	80	70
C	95	80
D	105	95
E	75	70
F	65	63
G	73	71
H	64	60
I	93	90
J	86	80
K	68	64
L	73	75

The *level of significance* is set at .05.

The *null hypothesis* states there is no significant difference between the subjects before and after treatment.

The *alternative hypothesis* states that there is a significant difference between the subjects before and after treatment.

Answer. Use the Wilcoxon test because the data is ranked. T = 2.5. The table value is 13. We reject the null hypothesis since the T value (2.5) is *less* than the table value (17). After the training program the secretaries improved their efficiency. (Chapter 6).

Program 12. A marriage counselor was interested in determining whether a woman's marital status was related to overall adjustment. From a population of college students she randomly selected six married women, nine single women and six divorcees. She had each woman rated on overall adjustment. The ratings follow.

Overall Adjustment Ratings

Married	Single	Divorced
20	4	15
25	10	24
32	34	30
40	26	18
28	28	5
43	22	2
	12	
	35	
	42	

The *level of significance* is set at .01.
The *null hypothesis* states there is no significant difference in the overall adjustment of the three groups.
The *alternative hypothesis* states there is a significant difference in the overall adjustment of the three groups.

Answer. The Kruskal-Wallis test is used. H = 4.69. Use Table A with a df of 2. You have a table value of 9.21. You fail to reject the null hypothesis because your value for H is not equal to or greater than your table value. There appears to be no significant difference between the three groups in overall adjustment. (Chapter 6)

Problem 13. A social worker wanted to know if children who are adopted changed in their personal happiness. A random sample of 70 children who were just going to be adopted were polled. They were asked to answer yes if they were happy and no if they were not happy. Six months after these children were adopted, they were polled again and asked the very same question. The children's responses are recorded in four possible ways, as shown in the table which follows.

		After Adoption	
		Unhappy	Happy
Before Adoption	Happy	8	16
	Unhappy	6	40

The *level of significance* is set at .05.

The *null hypothesis* states that children's attitudes do not differ significantly.

The *alternative hypothesis* states that childrens' attitudes are different.

Answer. We use the McNemar test. $\chi^2 = 20.02$. The table value is 3.84. We reject the null hypothesis, because the chi square value 20.02 is greater than your table value of 3.84. In this case, children's attitudes did change. (Chapter 5)

Problem 14. A calculus teacher wanted to find out the best way to demonstrate a proof to his advanced 12th grade class. He had three possible ways of proving a formula and he felt the students would not like them the same. He explained each of these three proofs to his class of 12 students and asked each student to write down 1 if he understood the proof and 0 if he didn't. The data from his study is shown in the following table.

	Methods		
Student	I	II	III
A	1	1	1
B	1	0	0
C	1	0	0
D	1	0	0
E	0	0	1
F	1	0	0
G	0	0	0
H	1	0	1
I	1	0	1
J	1	0	1
K	1	0	1
L	1	1	0

The *level of significance* is set at .05.

The *null hypothesis* states there is no significant difference among the three groups.

The *alternative hypothesis* states the three groups are different.

Answer. We use the Cochran Q test. Q = 9.60. The table value is 5.99 (df = 2). We reject the null hypothesis because the value for Q(9.60) is greater than the table value (5.99). In this problem there is a significant difference; the three methods are certainly different. The three groups are related because we tried *each* method on *each* member of the group. (Chapter 5)

Problem 15. An experimenter wanted to test the effectiveness of four seed treatments on soybean seeds. He classified four different plots of land so that soil fertility and growing conditions were uniform. He wanted the plots to be close together in giving similar yields. Each plot of land was then assigned to a treatment. At the end of the experiment he recorded the number of plants which failed to emerge out of 75 planted. The results follow.

PLANT FAILURES

Block	Treatment			
	1	2	3	4
A	6	8	9	11
B	2	4	5	9
C	3	8	8	7
D	6	4	8	7

The *level of significance* is set at .05.
The *null hypothesis* states there is no significant difference among the four treatments.
The *alternative hypothesis* states the four treatments are different.

Answer. We use the Randomized Block Design, One-Way Analysis of Variance. $F = 5.24$. The table value is 3.86. The between-groups $df = 3$, and the residual (error) $df = 9$. We reject the null hypothesis. The F ratio (5.24) is greater than the table value (3.86). The four treatments are significantly different. (Chapter 7)

Problem 16. Goldman, a supermarket specialist, wants to discern if the choice of cantaloupes is affected by their color. To find out, he arranges to have cantaloupes placed before twenty-five randomly selected customers. The cantaloupes range from very dark (ranked 1) to extremely light (ranked 5). Each customer is offered a choice among five different piles. The customers' selections follow.

Customers' Cantaloupe Preferences

Rank	Observed Choices
1	10
2	8
3	4
4	2
5	1

The *level of significance* is set at .01.
The *null hypothesis* states there is no significant difference in the preferences of the customers.

The *alternative hypothesis* states there is a difference in the customers' preferences. A two-tail test is used.

Answer. The Kolmogorov-Smirnov Test is used. D = .32. The table value is .32. We reject the null hypothesis since the value for D(.32) is equal to the table value (.32). We conclude that the color of cantaloupes does seem to affect customers' selections. (Chapter 6)

Problem 17. A college researcher randomly selected ten college students to serve as a mock jury. They were given a description of the case and told to evaluate the defendent's guilt or innocence on an 8-point scale, one being completely innocent and eight being completely guilty. At two points during the trial they were then asked to re-evaluate the defendants using the same rating system. The results follow.

Mock Trial Ratings

Student	Measurement I Scores	Measurement II Scores	Measurement III Scores
A	5	4	2
B	8	2	1
C	7	3	2
D	3	2	1
E	5	5	4
F	2	3	3
G	7	6	5
H	8	7	5
I	3	5	6
J	8	3	1

The *level of significance* is set at .05.
The *null hypothesis* states that there is no significant difference among the three evaluations.
The *alternative hypothesis* states that there is a significant difference among the three evaluations.

Answer. We use the Friedman test. $\chi r^2 = 254.48$. The table value is 5.99. You use Table A with a df = 2. Since your χr^2 value (254.48) is greater than your table value (5.99), you reject the null hypothesis. There is a significant difference among the three evaluations. (Chapter 6)

Problem 18. A psychometrist is concerned about students' knowledge in statistics. On a standardized test in advanced statistics, the mean score was shown to be 130 last year. A class of

15 students takes this test, and the resulting mean is 100. The psychometrist wants to test the hypothesis that the average score for this present class on the test is different. The present class of student test scores follow.

Student Scores

150
95
80
90
75
125
95
98
100
103
97
120
92
91
89

The *level of significance* is set at .05.
The *null hypothesis* states that the mean of the population is 130.
The *alternative hypothesis* states that the mean of the population is not 130. A two-tail test is used.

Answer. We use the t(I) test. t = −6.16. The table value is 2.145. We reject the null hypothesis, since the t value (−6.16) is greater than the table value (2.145). The mean of the sample is significantly different from the assumed population mean of 130. (Chapter 7)

Problem 19. The head of a police academy wants to determine the relationship between scores on the entrance exam and the gender of his new cadets. Ten students are randomly chosen and their exam scores and gender are recorded. The data follows.

Police Academy Data

Student	Exam Score Y	Gender
A	95	Female
B	90	Male
C	80	Female
D	70	Male

Police Academy Data

	Exam Score Y	Gender
E	88	Female
F	65	Male
G	74	Female
H	70	Male
I	97	Female
J	93	Male

The *level of significance* is set at .05.

The *null hypothesis* states there is no relationship between gender and exam scores.

The *alternative hypothesis* states that exam scores and gender are related. A two-tail test is used.

Answer. Use the Point-Biserial technique, and be sure to indicate the gender as 1 or 0. $r_{pb} = .41$. Using Table G, a df of 8 you find a table value of 2.306. You have a t value of 1.27. You fail to reject the null hypothesis in this case because the t value (1.27) is not equal to or greater than the table value of (2.306). There appears to be no relationship between the two variables. (Chapter 9)

Problem 20. When you fry chicken, it absorbs cooking oil in varying degrees. Gloria wants to learn if the amount of cooking oil absorption of chicken depends upon the type of cooking oil used. She wishes to test four different brands of cooking oil. For each type of cooking oil, five equivalent fryer chickens are prepared. The data in the following table gives the ounces of oil that are absorbed by the chickens.

Ounces of Oil Absorbed by Chickens

Cooking Oil	Group A	Group B	Group C	Group D
Brand V	1	2.5	1	4
Brand W	3.5	2	1	3
Brand X	2	1	3	4
Brand Y	2	2	4	4
Brand Z	3	2	3	4

The *level of significance* is set at .05.

The *null hypothesis* states there is no significant difference among the four groups of chickens.

The *alternative hypothesis* states the four groups of chicken are different.

Answer. We use the one-way analysis of variance-randomized group design. F = 4.23. The table value is 3.24. The Between df = 3, the Within df = 16. We reject the null hypothesis. The F value (4.23) is greater than your table value (3.24). There is a difference among the four groups. (Chapter 7)

Problem 20A. Looking at Problem 20, you are to use the Scheffé test and report whether Group A and Group D had differences which are significant. The data for the two groups follows.

Group A	Group D
1	4
3.5	3
2	4
2	4
3	4

Answer. F = 2.31. (The MS_w = .8125) Group A and Group D are not significant, since the F value (2.31) is not equal to or greater than the table value (3.24). (Chapter 7)

Problem 20B. Using the Scheffé test again on problem 20, report whether Group B and Group D had differences that were significant. The data follows.

Group B	Group D
2.5	4
2	3
1	4
2	4
2	4

Answer. F = 3.71. We find that Group B and Group D are significantly different, because the F value (3.71) is greater than the table value 3.24.

Problem 20C. Finally, use the Scheffé test again on problem 20 and report whether Group A and Group C had differences that were significant. The data follows.

Group A	Group C
1	1
3.5	1
2	3
2	4
3	3

Answer. F = .01. We find that Group A and Group C are not significantly different, because the F value (.01) is not equal to or greater than the table value (3.24).

Problem 21. A statistics test was administered to a class of 10. The scores follow. Calculate the mean, standard error of mean

Class Scores

40
50
65
89
95
100
75
70
77
72

and the standard deviation.

Answer. The mean is equal to 73.30, the standard deviation is equal to 17.83, and the standard error of the mean is equal to 5.64. (Chapter 2)

Problem 22. A history test was administered to a class of 15. The scores follow. Calculate the mean, standard error of the mean, and standard deviation.

History Test Scores

95
80
60
20
10
88
75
33
45
89
99
97
86
55
77

Answer. The mean is equal to 67.27, the standard deviation is equal to 27.83 and the standard error of the mean is equal to 7.19.

Problem 23. Four parole board members rank nine convicts on whether they are ready to be paroled. The four independent rankings that the parole board members gave are shown in the following table.

Parole Board Members	Convicts								
	a	b	c	d	e	f	g	h	i
W	9	8	7	3	4	5	6	1	2
X	7	8	9	1	2	4	3	5	6
Y	9	8	3	4	1	5	7	6	2
Z	4	5	3	1	2	9	6	7	8

The *level of significance* is set at .05.

The *null hypothesis* states there is no relationship among the rankings of the four parole board members.

The *alternative hypothesis* states there is agreement in the rankings of the four parole board members.

Answer. We use the Kendall Concordance Coefficient of Correlation. W = .4437500. Using Table A with a df of 8 you find a table value of 15.51. You have a χ^2 equal to 14.20. You fail to reject the null hypothesis, since the chi square value (14.20) is not equal to or greater than the table value (15.51). (Chapter 9)

Problem 24. Compute the Kendall tau for this problem. Junior and Senior College students rate eight instructors on knowledge of subject matter. The scores are tabulated and shown in the following table.

Instructor	Junior Students	Senior Students
1	40	27
2	45	50
3	30	30
4	35	28
5	38	43
6	22	20
7	25	22
8	15	26
9	11	13
10	9	8
11	5	6

The *level of significance* is set at .01.

The *null hypothesis* states there is no relationship between the rankings of the two groups of students.

The *alternative hypothesis* states that there is a positive relationship between the rankings of the two groups of students. A one-tail test is used. Note that before you solve this problem, you must turn these scores into ranks.

Answer. T = +.7818181. We find a z equal to 3.35 and its probability of occurring is smaller than .0005. We reject the null hypothesis because the probability .0005 is less than the level of significance of .01. There appears to be an agreement in the rankings of the two classes. (Chapter 9)

Problem 25. A gambler wanted to see if a die he had just bought was honest. He rolled this die sixty times. The table follows with the results.

Die Rolled Sixty Times

	1	2	3	4	5	6
Observations	20	5	6	7	18	4

The *level of significance* is set at .05.

The *null hypothesis* states there is no significant difference between the expected and observed frequencies for the die rolls.

The *alternative hypothesis* states the expected and observed frequencies for the die rolls are different.

Answer. We use the $\chi^2(I)$ test. Note, you must figure the expected frequencies for this problem. $\chi^2 = 25$. The table value is 11.07. Since the chi square value (25) is greater than the table value 11.07, reject the null hypothesis. The faces of the die are not all the same (Chapter 5)

Problem 26. A group of 100 salesmen were classified with respect to appearance and income level. Three categories of appearance were used: attractive, average and unattractive. Three categories of income were used: high, medium and low. This classification produced the following table.

	Income		
Appearance	High	Medium	Low
Attractive	20	11	6
Average	13	9	8
Unattractive	8	7	18

The *level of significance* is set at .01.

The *null hypothesis* states that there is no significance among the three groups.

The *alternative hypothesis* states that the three groups are different.

Answer. We use the $\chi^2(II)$. $\chi^2(II) = 12.85$. The table value is 13.28 with a df of 4. We fail to reject the null hypothesis, because the chi square value (12.85) is not equal to or greater than the table value (13.28). The groups appear to be the same (Chapter 5)

Problem 27. The correlation between grades in high school and a test of mechanical skill is .4 (r_{12}). The correlation between high school grades and a test of eye and hand coordination is .3 (r_{13}). The correlation between the test of mechanical skill and the eye and hand coordination test is a .8 (r_{23}). Find the multiple correlation between high school grades and the combined effects of mechanical skill and eye and hand coordination scores.

Answer. The multiple correlation between the high school grades and the combined effects of mechanical skill and eye and hand coordination is .40, $r_{1.23} = .40$. (Chapter 8)

Problem 28. Suppose we now say the correlation between grades in high school and a test of mechanical skill is .20 (r_{12}). The correlation between high school grades and a test of eye and hand coordination is .11 (r_{13}). The correlation between the test of mechanical skill and the eye and hand coordination test is .90 (r_{23}). Find the multiple correlation between high school grades and the combined effects of mechanical skill and eye and hand coordination scores.

Answer. The multiple correlation between the high school grades and the combined effects of mechanical skill and eye and hand coordination is now .26, $r_{1.23} = .26$.

Problem 29. A college instructor wanted to see the effect of programmed instruction on students in an accounting course. A group of students were given a test in accounting. They then had programmed instruction for six months and were readministered an alternate form of the same test. The results are shown:

Student	Accounting Test I	Test Scores Test II
A	50	60
B	75	90
C	80	99
D	45	75
E	20	88
G	50	60
H	70	85
I	66	73
J	33	45
K	78	87

The *level of significance* is set at .05.

The *null hypothesis* states that there is no significant difference between the two groups of scores.

The *alternative hypothesis* states that the group taking the test for the second time will improve. A one-tail test is used.

Answer. We use a t(III) test. Remember, the same individuals are being matched with themselves. t(III) = −3.37. The table value is 1.860. We reject the null hypothesis. The t value (−3.37) is greater than the table value (1.860). There is a significant difference. The group taking the test for the second time improved. (Chapter 7)

Problem 30. Paul took a random sample of 66 people and recorded whether they were for or against the ERA. He then showed them a film favorable to the ERA. One month later he recorded their position on this issue. These individuals' responses are recorded in the table that follows.

Before the Film	After Film	
	For ERA	Against ERA
Against ERA	30	10
For ERA	16	10

The *level of significance* is .05.

The *null hypothesis* states that people's attitudes toward the ERA remains the same.

The *alternative hypothesis* states that the poeple's attitude toward the ERA is different.

Answer. We use the McNemar test. $\chi^2 = 9.03$. The table value is 3.84 with a df of 1. We reject the null hypothesis, since the value for the McNemar 9.03 is greater than the table value (3.84). The people's attitude toward the ERA is different. (Chapter 5)

Problem 31. A researcher wants to determine the relationship between stanine scores on a math test and whether or not a student passed his language proficiency exam. Ten students were chosen at random and their math stanine scores and their language proficiency status was recorded. The results are shown in the following table.

Student	Math Stanine	Language Proficiency Status
A	8	Pass
B	3	Fail
C	1	Fail
D	5	Pass
E	6	Fail
F	7	Pass
G	2	Fail
H	8	Pass

Student	Math Stanine	Language Proficiency Status
I	9	Pass
J	6	Fail

The *level of significance* is set at .05.

The *null hypothesis* states that there is no relationship between Language Proficiency Status and Math test scores.

The *alternative hypothesis* states that math test scores and language proficiency status are related. At two-tail test is used.

Answer. A point-biserial test is used; r_{pb} = .74. Using Table G with a df of 8, you find a table value equal to 2.228. You have a t value of 4.21. You reject the null hypothesis, since the t value of 4.21 is greater than your table value of 2.228. There appears to be a relationship between language proficiency status and scores on the math test. (Chapter 9)

Problem 32. A dog expert wants to find out if vitamins have an effect on dogs. She obtains ten pairs of identical dogs to serve as subjects. These dogs are matched on the basis of size, appearance, color, type, and vitality. At random, one dog from each set is assigned to a group. One group of dogs is then feed special vitamins along with the prescribed diet for two months while the other group is just given the prescribed diet. At the end of this period, a panel of judges ranked all the dogs on the basis of their appearance and vitality. The results are shown in the following table.

Pair	Group A Vitamins	Group B No Vitamins
A	8	10
B	9	6
C	8	7
D	8	3
E	10	3
F	4	5
G	2	9
H	4	7
I	6	6
J	9	8

The *level of significance* is set at .01.

The *null hypothesis* states there is no significant difference between Group A and Group B.

The *alternative hypothesis* states that Group A will do better than Group B. A one-tail test is used.

Answer. We use the Wilcoxon test. T = 20. The table value equals 3, remember this is a one-tail test. We fail to reject the null hypothesis, since the T value (20) is not *equal* to or *less* than the table value (3). The dogs for both groups are the same. (Chapter 6)

Problem 33. Army trainees were given a manual dexterity test and the Army Aptitude test to determine their skills. The men in charge of the program wanted to determine how close these aptitude scores were related to the scores on the manual dexterity test. The trainees' scores on both tests are recorded in the following table.

Trainee	Aptitude Test	Manual Dexterity Test
A	45	80
B	56	50
C	99	79
D	70	80
E	60	67
F	55	30
G	80	90
H	55	60
I	100	90
J	80	70

The *level of significance* is set at .05.

The *null hypothesis* states there is no correlation between these two variables.

The *alternative hypothesis* states there is a relationship between the two variables. A two-tail test is used.

Answer. The Pearson Product Moment Correlation is used. r = .13. You use Table L. The degrees of freedom equal 8. Your table value is .6319. Since your r of .13 is not equal to or greater than your table value of .6319, we fail to reject the null hypothesis. There appears to be no relationship between the two variables. (Chapter 8)

Problem 34. A history teacher wanted to determine whether movies or lectures were more effective in teaching students about the American Revolution. She randomly selects two groups of students for the experiment. Group I is to receive the usual lecture hall method of instruction, while Group II is to receive a series of films on the American Revolution. At the end of a period of six months, the two groups are given a standardized history exam. The scores for the two groups follow.

Group I	Group II
10	40
67	80
99	35
70	80
90	95
88	85
45	55
60	77
78	74
85	89

The *level of significance* is set at .05.

The *null hypothesis* states there is no significant difference between the two groups.

The *alternative hypothesis* states that the two groups are different. A two-tail test is used.

Answer. We use a t(II) test. t = −.17. The table value is 2.101. We fail to reject the null hypothesis, since the t value (1−.17) is not equal to or greater than the table value (2.101). There is no significant difference between the two teaching methods. (Chapter 7)

Problem 35. A psychologist wants to see if there is a correlation between passing or failing a calculus course and being anxious or not about taking the test. He randomly selects 110 students and classifies them as a pass or fail on the calculus test and anxious or not anxious before they take the exam. The data are recorded in the following table.

Emotional State	Pass	Fail
Anxious	50	10
Not Anxious	30	20

The *level of significance* is set at .01.

The *null hypothesis* states there is no relationship between the two variables.

The *alternative hypothesis* states that there is a relationship between passing or failing calculus and emotional state.

Answer. The Contingency Coefficient of Correlation is used. We reject the null hypothesis. $\chi^2 = 6.64$. The table value is 6.64. Since the chi square value of 6.64 is equal to the table value of 6.64, we can reject the null hypothesis. There appears to be a relationship between emotional state and passing or failing calculus. The relationship between the two variables is not 0. The

C value is .24; this gives you the magnitude of the association between the variables. (Chapter 8).

Problem 36. A manufacturer of jeans showed his latest five designs (A, B, C, D, E) to twenty buyers from different retail stores. When an order is placed it is designated by an x the results follow.

Orders for Jeans

Buyer	A	B	C	D	E
1	x	x		x	
2	x		x		x
3	x		x	x	
4	x	x			x
5		x	x		
6	x				x
7	x	x		x	
8	x	x			
9			x	x	x
10					x
11			x		
12	x				
13	x		x		
14	x				x
15	x				
16	x	x	x	x	
17	x				
18	x				
19		x	x		
20	x	x			

The *level of significance* is set at .05.

The *null hypothesis* states there is no significant difference among the five jean designs.

The *alternative hypothesis* states the demand for each of the five styles of jeans is significantly different.

Answer. We use the Cochran Q test. $Q = 11.55$. The table value is 9.49. ($df = 4$) We reject the null hypothesis in this case since the value for Q(11.55) is greater than the table value (9.49). In this problem there was a significant difference in the demand for the five styles of jeans. When you do your Cochran Q, check to see that you only assigned values of 0 or 1. (Chapter 5).

Problem 37. The Martinez family owned a drugstore that carried imported chocolate bars. They wanted to know if there was a significant difference in the customer's preference for the candy. The owner was in favor of the less expensive brands of candy, but her husband disagreed. To settle the question, they gave ten randomly selected customers one of each type of chocolate bar to try and asked them to rate them, 1 for the best, 2 second best, 3 third best and 4 the least preferred. The results follow.

Adult Rater	Candy Bar $1.25	Candy Bar $1.20	Candy Bar $1.15	Candy Bar $1.10
A	1	3	2	4
B	2	3	1	4
C	4	3	2	1
D	3	1	2	4
E	1	2	4	3
F	1	2	3	4
G	4	3	2	1
H	2	1	3	4
I	3	2	1	4
J	4	3	2	1

The *level of significance* is set at .05.

The *null hypothesis* states there is no significant difference among the candy bars.

The *alternative hypothesis* states that the candy bars are different.

Answer. We use the Friedman test. $\chi r^2 = 2.28$. The table value is 9.49. You use Table A with df = 4. Since your χr^2 value (2.28) is not equal to or greater than your table value (9.49) we fail to reject the null hypothesis. There is no significant difference among the candy bars. (Chapter 6).

Problem 38. Three independent groups of subjects are rated on their social presence by a panel of judges. Each individual is assigned a social presence score. The three independent groups come from different social backgrounds, but our judges are not aware of this fact. The table that follows shows the results of the rating.

Background A	Background B	Background C
30	26	40
35	40	50
20	37	44
10	15	45
25	17	47

The *level of significance* is set at .05.
The *null hypothesis* states there is no significant difference among the three groups.
The *alternative hypothesis* states there is a difference.

Answer. We use the Kruskal-Wallis test. H = 9.16. Use Table G with column size 5, 5, 5. The probability of 9.16 occurring is less than .009. Since the probability is less than your level of significance (.05) reject the null hypothesis. There was a significant difference among the three groups. (Chapter 6).

Problem 39. A Nielsen pollster in a small New England town was interested in the type of sporting events that adults in the 25 to 50 age bracket liked to watch on television. He took a random sample of 200 adults in this age bracket and asked them whether they preferred baseball, football, basketball, hockey, or tennis on television. The results of this poll are shown in the following table.

Baseball	Football	Basketball	Hockey	Tennis
30	70	45	25	30

The *level of significance* is set at .01.
The *null hypothesis* states there is no significant difference between the expected and observed frequencies.
The *alternative hypothesis* states they are different.

Answer. We use the chi square (I) test. $\chi^2 = 33.75$. The table value is 13.28 with a df of 4. Since your chi square value (33.75) is greater than your table value (13.28), we reject the null hypothesis. The adults do not all feel the same about sporting events. (Chapter 5).

Problem 40. A computer specialist wanted to compare two approaches for training computer programmers. One approach used exclusively oral techniques, while the other used only written tests. The same program instructional material was used with both approaches. He randomly assigned fifteen students to work with oral approach and the remaining fifteen students to receive the written approach. At the end of six months, he tested both groups and recorded the results in the following table.

Group A Oral Approach	Group B Written Approach
20	40
30	50
50	60
70	80
85	94

| Group A | Group B |
Oral Approach	Written Approach
93	67
85	34
25	55
66	55
35	25
85	90
95	10
15	98
35	88
40	95

The *level of significance* is set at .05.

The *null hypothesis* states there is no significant difference between the two groups.

The *alternative hypothesis* states that the two groups are different. Use a two-tail test.

Answer. We use the t(II) test. t = −.72. The table value is 2.048 with a df of 28. We fail to reject the null hypothesis because the t value (−.72) is not equal to or greater than the table value (2.048). There appears to be no significant difference between the two groups.

Appendix A
Statistical Tables

Table A. Critical Values of Chi Square.

df	.99	.98	.95	.90	.80	.70	.50	.30	.20	.10	.05	.02	.01	.001
1	.0002	.0006	.0039	.016	.064	.15	.46	1.07	1.64	2.71	3.84	5.41	6.64	10.83
2	.02	.04	.10	.21	.45	.71	1.39	2.41	3.22	4.60	5.99	7.82	9.21	13.82
3	.12	.18	.35	.58	1.00	1.42	2.37	3.66	4.64	6.25	7.82	9.84	11.34	16.27
4	.30	.43	.71	1.06	1.65	2.20	3.36	4.88	5.99	7.78	9.49	11.67	13.28	18.47
5	.55	.75	1.14	1.61	2.34	3.00	4.35	6.06	7.29	9.24	11.07	13.39	15.09	20.52
6	.87	1.13	1.64	2.20	3.07	3.83	5.35	7.23	8.56	10.64	12.59	15.03	16.81	22.46
7	1.24	1.56	2.17	2.83	3.82	4.67	6.35	8.38	9.80	12.02	14.07	16.62	18.48	24.32
8	1.65	2.03	2.73	3.49	4.59	5.53	7.34	9.52	11.03	13.36	15.51	18.17	20.09	26.12
9	2.09	2.53	3.32	4.17	5.38	6.39	8.34	10.66	12.24	14.68	16.92	19.68	21.67	27.88
10	2.56	3.06	3.94	4.86	6.18	7.27	9.34	11.78	13.44	15.99	18.31	21.16	23.21	29.59
11	3.05	3.61	4.58	5.58	6.99	8.15	10.34	12.90	14.63	17.28	19.68	22.62	24.72	31.26
12	3.57	4.18	5.23	6.30	7.81	9.03	11.34	14.01	15.81	18.55	21.03	24.05	26.22	32.91
13	4.11	4.76	5.89	7.04	8.63	9.93	12.34	15.12	16.98	19.81	22.36	25.47	27.69	34.53
14	4.66	5.37	6.57	7.79	9.47	10.82	13.34	16.22	18.15	21.06	23.68	26.87	29.14	36.12
15	5.23	5.98	7.26	8.55	10.31	11.72	14.34	17.32	19.31	22.31	25.00	28.26	30.58	37.70

16	5.81	6.61	7.96	9.31	11.15	12.62	15.34	18.42	20.46	23.54	26.30	29.63	32.00	39.25
17	6.41	7.26	8.67	10.08	12.00	13.53	16.34	19.51	21.62	24.77	27.59	31.00	33.41	40.79
18	7.02	7.91	9.39	10.86	12.86	14.44	17.34	20.60	22.76	25.99	28.87	32.35	34.80	42.31
19	7.63	8.57	10.12	11.65	13.72	15.35	18.34	21.69	23.90	27.20	30.14	33.69	36.19	43.82
20	8.26	9.24	10.85	12.44	14.58	16.27	19.34	22.78	25.04	28.41	31.41	35.02	37.57	45.32
21	8.90	9.92	11.59	13.24	15.44	17.18	20.34	23.86	26.17	29.62	32.67	36.34	38.93	46.80
22	9.54	10.60	12.34	14.04	16.31	18.10	21.34	24.94	27.30	30.81	33.92	37.66	40.29	48.27
23	10.20	11.29	13.09	14.85	17.19	19.02	22.34	26.02	28.43	32.01	35.17	38.97	41.64	49.73
24	10.86	11.99	13.85	15.66	18.06	19.94	23.34	27.10	29.55	33.20	36.42	40.27	42.98	51.18
25	11.52	12.70	14.61	16.47	18.94	20.87	24.34	28.17	30.68	34.38	37.65	41.57	44.31	52.62
26	12.20	13.41	15.38	17.29	19.82	21.79	25.34	29.25	31.80	35.56	38.88	42.86	45.64	54.05
27	12.88	14.12	16.15	18.11	20.70	22.72	26.34	30.32	32.91	36.74	40.11	44.14	46.96	55.48
28	13.56	14.85	16.93	18.94	21.59	23.65	27.34	31.39	34.03	37.92	41.34	45.42	48.28	56.89
29	14.26	15.57	17.71	19.77	22.48	24.58	28.34	32.46	35.14	39.09	42.56	46.69	49.59	58.30
30	14.95	16.31	18.49	20.60	23.36	25.51	29.34	33.53	36.25	40.26	43.77	47.96	50.89	59.70

Table B. Probabilities Associated with Values as Small as the Observed Values of U in the Mann-Whitney Test When n_2 is Between 3 and 80.

$n_2 = 3$

U \ n_1	1	2	3
0	.250	.100	.050
1	.500	.200	.100
2	.750	.400	.200
3		.600	.350
4			.500
5			.650

$n_2 = 4$

U \ n_1	1	2	3	4
0	.200	.067	.028	.014
1	.400	.133	.057	.029
2	.600	.267	.114	.057
3		.400	.200	.100
4		.600	.314	.171
5			.429	.243
6			.571	.343
7				.443
8				.557

$n_2 = 5$

U \ n_1	1	2	3	4	5
0	.167	.047	.018	.008	.004
1	.333	.095	.036	.016	.008
2	.500	.190	.071	.032	.016
3	.667	.286	.125	.056	.028
4		.429	.196	.095	.048
5		.571	.286	.143	.075
6			.393	.206	.111
7			.500	.278	.155
8			.607	.365	.210
9				.452	.274
10				.548	.345
11					.421
12					.500
13					.579

$n_2 = 6$

U \ n_1	1	2	3	4	5	6
0	.143	.036	.012	.005	.002	.001
1	.286	.071	.024	.010	.004	.002
2	.428	.143	.048	.019	.009	.004
3	.571	.214	.083	.033	.015	.008
4		.321	.131	.057	.026	.013
5		.429	.190	.086	.041	.021
6		.571	.274	.129	.063	.032
7			.357	.176	.089	.047
8			.452	.238	.123	.066
9			.548	.305	.165	.090
10				.381	.214	.120
11				.457	.268	.155
12				.545	.331	.197
13					.396	.242
14					.465	.294
15					.535	.350
16						.409
17						.469
18						.531

U \ n₁	1	2	3	$n_2 = 7$ 4	5	6	7
0	.125	.028	.008	.003	.001	.001	.000
1	.250	.056	.017	.006	.003	.001	.001
2	.375	.111	.033	.012	.005	.002	.001
3	.500	.167	.058	.021	.009	.004	.002
4	.625	.250	.092	.036	.015	.007	.003
5		.333	.133	.055	.024	.011	.006
6		.444	.192	.082	.037	.017	.009
7		.556	.258	.115	.053	.026	.013
8			.333	.158	.074	.037	.019
9			.417	.206	.101	.051	.027
10			.500	.264	.134	.069	.036
11			.583	.324	.172	.090	.049
12				.394	.216	.117	.064
13				.464	.265	.147	.082
14				.538	.319	.183	.104
15					.378	.223	.130
16					.438	.267	.159
17					.500	.314	.191
18					.562	.365	.228
19						.418	.267
20						.473	.310
21						.527	.355
22							.402
23							.451
24							.500
25							.549

U \ n₁	1	2	3	4	$n_2 = 8$ 5	6	7	8	t	Normal
0	.111	.022	.006	.002	.001	.000	.000	.000	3.308	.001
1	.222	.044	.012	.004	.002	.001	.000	.000	3.203	.001
2	.333	.089	.024	.008	.003	.001	.001	.000	3.098	.001
3	.444	.133	.042	.014	.005	.002	.001	.001	2.993	.001
4	.556	.200	.067	.024	.009	.004	.002	.001	2.888	.002
5		.267	.097	.036	.015	.006	.003	.001	2.783	.003
6		.356	.139	.055	.023	.010	.005	.002	2.678	.004
7		.444	.188	.077	.033	.015	.007	.003	2.573	.005
8		.556	.248	.107	.047	.021	.010	.005	2.468	.007
9			.315	.141	.064	.030	.014	.007	2.363	.009
10			.387	.184	.085	.041	.020	.010	2.258	.012
11			.461	.230	.111	.054	.027	.014	2.153	.016
12			.539	.285	.142	.071	.036	.019	2.048	.020
13				.341	.177	.091	.047	.025	1.943	.026
14				.404	.217	.114	.060	.032	1.838	.033
15				.467	.262	.141	.076	.041	1.733	.041
16				.533	.311	.172	.095	.052	1.628	.052
17					.362	.207	.116	.065	1.523	.064
18					.416	.245	.140	.080	1.418	.078
19					.472	.286	.168	.097	1.313	.094
20					.528	.331	.198	.117	1.208	.113
21						.377	.232	.139	1.102	.135
22						.426	.268	.164	.998	.159
23						.475	.306	.191	.893	.185
24						.525	.347	.221	.788	.215
25							.389	.253	.683	.247
26							.433	.287	.578	.282
27							.478	.323	.473	.318
28							.522	.360	.368	.356
29								.399	.263	.396
30								.439	.158	.437
31								.480	.052	.481
32								.520		

Table C. Critical Values of D for the Kolmogorov-Smirnov Test.

Sample Size (N)	Significance Level				
	.20	.15	.10	.05	.01
1	.900	.925	.950	.975	.995
2	.684	.726	.776	.842	.929
3	.565	.597	.642	.708	.829
4	.494	.525	.564	.624	.734
5	.446	.474	.510	.563	.669
6	.410	.436	.470	.521	.618
7	.381	.405	.438	.486	.577
8	.358	.381	.411	.457	.543
9	.339	.360	.388	.432	.514
10	.322	.342	.368	.409	.486
11	.307	.326	.352	.391	.468
12	.295	.313	.338	.375	.450
13	.284	.302	.325	.361	.433
14	.274	.292	.314	.349	.418
15	.266	.283	.304	.338	.404
16	.258	.274	.295	.328	.391
17	.250	.266	.286	.318	.380
18	.244	.259	.278	.309	.370
19	.237	.252	.272	.301	.361
20	.231	.246	.264	.294	.352
25	.21	.22	.24	.264	.32
30	.19	.20	.22	.242	.29
35	.18	.19	.21	.23	.27
Over 35	$\dfrac{1.07}{\sqrt{N}}$	$\dfrac{1.14}{\sqrt{N}}$	$\dfrac{1.22}{\sqrt{N}}$	$\dfrac{1.36}{\sqrt{N}}$	$\dfrac{1.63}{\sqrt{N}}$

Table D. Probabilities Associated with Values as Extreme as the Observed Values of z in the Normal Distribution.

The body of the table gives one-tailed probabilities under H_0 of z. The left-hand marginal column gives various values of z to one decimal place. The top row gives various values to the second decimal place. Thus, for example, the one-tailed p of $z \geq .11$ or $z \leq -.11$ is $p = .4562$.

z	.00	.01	.02	.03	.04	.05	.06	.07	.08	.09
.0	.5000	.4960	.4920	.4880	.4840	.4801	.4761	.4721	.4681	.4641
.1	.4602	.4562	.4522	.4483	.4443	.4404	.4364	.4325	.4286	.4247
.2	.4207	.4168	.4129	.4090	.4052	.4013	.3974	.3936	.3897	.3859
.3	.3821	.3783	.3745	.3707	.3669	.3632	.3594	.3557	.3520	.3483
.4	.3446	.3409	.3372	.3336	.3300	.3264	.3228	.3192	.3156	.3121
.5	.3085	.3050	.3015	.2981	.2946	.2912	.2877	.2843	.2810	.2776
.6	.2743	.2709	.2676	.2643	.2611	.2578	.2546	.2514	.2483	.2451
.7	.2420	.2389	.2358	.2327	.2296	.2266	.2236	.2206	.2177	.2148
.8	.2119	.2090	.2061	.2033	.2005	.1977	.1949	.1922	.1894	.1867
.9	.1841	.1814	.1788	.1762	.1736	.1711	.1685	.1660	.1635	.1611
1.0	.1587	.1562	.1539	.1515	.1492	.1469	.1446	.1423	.1401	.1379
1.1	.1357	.1335	.1314	.1292	.1271	.1251	.1230	.1210	.1190	.1170
1.2	.1151	.1131	.1112	.1093	.1075	.1056	.1038	.1020	.1003	.0985
1.3	.0968	.0951	.0934	.0918	.0901	.0885	.0869	.0853	.0838	.0823
1.4	.0808	.0793	.0778	.0764	.0749	.0735	.0721	.0708	.0694	.0681
1.5	.0668	.0655	.0643	.0630	.0618	.0606	.0594	.0582	.0571	.0559
1.6	.0548	.0537	.0526	.0516	.0505	.0495	.0485	.0475	.0465	.0455
1.7	.0446	.0436	.0427	.0418	.0409	.0401	.0392	.0384	.0375	.0367
1.8	.0359	.0351	.0344	.0336	.0329	.0322	.0314	.0307	.0301	.0294
1.9	.0287	.0281	.0274	.0268	.0262	.0256	.0250	.0244	.0239	.0233
2.0	.0228	.0222	.0217	.0212	.0207	.0202	.0197	.0192	.0188	.0183
2.1	.0179	.0174	.0170	.0166	.0162	.0158	.0154	.0150	.0146	.0143
2.2	.0139	.0136	.0132	.0129	.0125	.0122	.0119	.0116	.0113	.0110
2.3	.0107	.0104	.0102	.0099	.0096	.0094	.0091	.0089	.0087	.0084
2.4	.0082	.0080	.0078	.0075	.0073	.0071	.0069	.0068	.0066	.0064
2.5	.0062	.0060	.0059	.0057	.0055	.0054	.0052	.0051	.0049	.0048
2.6	.0047	.0045	.0044	.0043	.0041	.0040	.0039	.0038	.0037	.0036
2.7	.0035	.0034	.0033	.0032	.0031	.0030	.0029	.0028	.0027	.0026
2.8	.0026	.0025	.0024	.0023	.0023	.0022	.0021	.0021	.0020	.0019
2.9	.0019	.0018	.0018	.0017	.0016	.0016	.0015	.0015	.0014	.0014
3.0	.0013	.0013	.0013	.0012	.0012	.0011	.0011	.0011	.0010	.0010
3.1	.0010	.0009	.0009	.0009	.0008	.0008	.0008	.0008	.0007	.0007
3.2	.0007									
3.3	.0005									
3.4	.0003									
3.5	.00023									
3.6	.00016									
3.7	.00011									
3.8	.00007									
3.9	.00005									
4.0	.00003									

Table E. Critical Values of U in the Mann-Whitney Test When N₂ is Between 9 and 20.

Critical Values of U for a One-tailed Test at $\alpha = .01$ or for a Two-tailed Test at $\alpha = .02$

n_1 \ n_2	9	10	11	12	13	14	15	16	17	18	19	20
1												
2												
3									0	0	0	0
4		0	0	0	1	1	1	2	2	3	3	3
5	1	1	2	2	3	3	4	5	5	6	7	7
6	2	3	4	4	5	6	7	8	9	10	11	12
7	3	5	6	7	8	9	10	11	13	14	15	16
8	5	6	8	9	11	12	14	15	17	18	20	21
9	7	8	10	12	14	15	17	19	21	23	25	26
10	8	10	12	14	17	19	21	23	25	27	29	32
11	10	12	15	17	20	22	24	27	29	32	34	37
12	12	14	17	20	23	25	28	31	34	37	40	42
13	14	17	20	23	26	29	32	35	38	42	45	48
14	15	19	22	25	29	32	36	39	43	46	50	54
15	17	21	24	28	32	36	40	43	47	51	55	59
16	19	23	27	31	35	39	43	48	52	56	60	65
17	21	25	29	34	38	43	47	52	57	61	66	70
18	23	27	32	37	42	46	51	56	61	66	71	76
19	25	29	34	40	45	50	55	60	66	71	77	82
20	26	32	37	42	48	54	59	65	70	76	82	88

Critical Values of U for a One-tailed Test at $\alpha = .01$ or for a Two-tailed Test at $\alpha = .02$

n_1 \ n_2	9	10	11	12	13	14	15	16	17	18	19	20
1												
2					0	0	0	0	0	0	1	1
3	1	1	1	2	2	2	3	3	4	4	4	5
4	3	3	4	5	5	6	7	7	8	9	9	10
5	5	6	7	8	9	10	11	12	13	14	15	16
6	7	8	9	11	12	13	15	16	18	19	20	22
7	9	11	12	14	16	17	19	21	23	24	26	28
8	11	13	15	17	20	22	24	26	28	30	32	34
9	14	16	18	21	23	26	28	31	33	36	38	40
10	16	19	22	24	27	30	33	36	38	41	44	47
11	18	22	25	28	31	34	37	41	44	47	50	53
12	21	24	28	31	35	38	42	46	49	53	56	60
13	23	27	31	35	39	43	47	51	55	59	63	67
14	26	30	34	38	43	47	51	56	60	65	69	73
15	28	33	37	42	47	51	56	61	66	70	75	80
16	31	36	41	46	51	56	61	66	71	76	82	87
17	33	38	44	49	55	60	66	71	77	82	88	93
18	36	41	47	53	59	65	70	76	82	88	94	100
19	38	44	50	56	63	69	75	82	88	94	101	107
20	40	47	53	60	67	73	80	87	93	100	107	114

Critical Values of U for a One-tailed Test at $\alpha = .05$ or for a Two-tailed Test at $\alpha = .10$

n_2 \ n_1	9	10	11	12	13	14	15	16	17	18	19	20
1											0	0
2	1	1	1	2	2	2	3	3	3	4	4	4
3	3	4	5	5	6	7	7	8	9	9	10	11
4	6	7	8	9	10	11	12	14	15	16	17	18
5	9	11	12	13	15	16	18	19	20	22	23	25
6	12	14	16	17	19	21	23	25	26	28	30	32
7	15	17	19	21	24	26	28	30	33	35	37	39
8	18	20	23	26	28	31	33	36	39	41	44	47
9	21	24	27	30	33	36	39	42	45	48	51	54
10	24	27	31	34	37	41	44	48	51	55	58	62
11	27	31	34	38	42	46	50	54	57	61	65	69
12	30	34	38	42	47	51	55	60	64	68	72	77
13	33	37	42	47	51	56	61	65	70	75	80	84
14	36	41	46	51	56	61	66	71	77	82	87	92
15	39	44	50	55	61	66	72	77	83	88	94	100
16	42	48	54	60	65	71	77	83	89	95	101	107
17	45	51	57	64	70	77	83	89	96	102	109	115
18	48	55	61	68	75	82	88	95	102	109	116	123
19	51	58	65	72	80	87	94	101	109	116	123	130
20	54	62	69	77	84	92	100	107	115	123	130	138

Critical Values of U for a One-tailed Test at $\alpha = .025$ or for a Two-tailed Test at $\alpha = .05$

n_2 \ n_1	9	10	11	12	13	14	15	16	17	18	19	20
1												
2	0	0	0	1	1	1	1	1	2	2	2	2
3	2	3	3	4	4	5	5	6	6	7	7	8
4	4	5	6	7	8	9	10	11	11	12	13	13
5	7	8	9	11	12	13	14	15	17	18	19	20
6	10	11	13	14	16	17	19	21	22	24	25	27
7	12	14	16	18	20	22	24	26	28	30	32	34
8	15	17	19	22	24	26	29	31	34	36	38	41
9	17	20	23	26	28	31	34	37	39	42	45	48
10	20	23	26	29	33	36	39	42	45	48	52	55
11	23	26	30	33	37	40	44	47	51	55	58	62
12	26	29	33	37	41	45	49	53	57	61	65	69
13	28	33	37	41	45	50	54	59	63	67	72	76
14	31	36	40	45	50	55	59	64	67	74	78	83
15	34	39	44	49	54	59	64	70	75	80	85	90
16	37	42	47	53	59	64	70	75	81	86	92	98
17	39	45	51	57	63	67	75	81	87	93	99	105
18	42	48	55	61	67	74	80	86	93	99	106	112
19	45	52	58	65	72	78	85	92	99	106	113	119
20	48	55	62	69	76	83	90	98	105	112	119	127

Table F. Critical Values of T for the Wilcoxon Test.

N	Level of significance for one-tail test			
	.05	.025	.01	.005
	Level of significance for two-tail test			
	.10	.05	.02	.01
6	2	1	—	—
7	4	2	0	—
8	6	4	2	0
9	8	6	3	2
10	11	8	5	3
11	14	11	7	5
12	17	14	10	7
13	21	17	13	10
14	26	21	16	13
15	30	25	20	16
16	36	30	24	19
17	41	35	28	23
18	47	40	33	28
19	54	46	38	32
20	60	52	43	37
21	68	59	49	43
22	75	66	56	49
23	83	73	62	55
24	92	81	69	61
25	101	90	77	68

Table G. Critical Values of t.

	Level of significance for a one-tail test				
	.05	.025	.01	.005	.0005
	Level of significance for a two-tail test				
df	.10	.05	.02	.01	.001
1	6·314	12·706	31·821	63·657	636·619
2	2·920	4·303	6·965	9·925	31·598
3	2·353	3·182	4·541	5·841	12·924
4	2·132	2·776	3·747	4·604	8·610
5	2·015	2·571	3·365	4·032	6·869
6	1·943	2·447	3·143	3·707	5·959
7	1·895	2·365	2·998	3·499	5·408
8	1·860	2·306	2·896	3·355	5·041
9	1·833	2·262	2·821	3·250	4·781
10	1·812	2·228	2·764	3·169	4·587
11	1·796	2·201	2·718	3·106	4·437
12	1·782	2·179	2·681	3·055	4·318
13	1·771	2·160	2·650	3·012	4·221
14	1·761	2·145	2·624	2·977	4·140
15	1·753	2·131	2·602	2·947	4·073
16	1·746	2·120	2·583	2·921	4·015
17	1·740	2·110	2·567	2·898	3·965
18	1·734	2·101	2·552	2·878	3·922
19	1·729	2·093	2·539	2·861	3·883
20	1·725	2·086	2·528	2·845	3·850
21	1·721	2·080	2·518	2·831	3·819
22	1·717	2·074	2·508	2·819	3·792
23	1·714	2·069	2·500	2·807	3·767
24	1·711	2·064	2·492	2·797	3·745
25	1·708	2·060	2·485	2·787	3·725
26	1·706	2·056	2·479	2·779	3·707
27	1·703	2·052	2·473	2·771	3·690
28	1·701	2·048	2·467	2·763	3·674
29	1·699	2·045	2·462	2·756	3·659
30	1·697	2·042	2·457	2·750	3·646
40	1·684	2·021	2·423	2·704	3·551
60	1·671	2·000	2·390	2·660	3·460
120	1·658	1·980	2·358	2·617	3·373
∞	1·645	1·960	2·326	2·576	3·291

Table H. Probabilities Associated with Values as Large as the Observed Values of H in the Kruskal-Wallis Test.

Sample sizes			H	p	Sample sizes			H	p
n_1	n_2	n_3			n_1	n_2	n_3		
2	1	1	2.7000	.500	4	3	2	6.4444	.008
								6.3000	.011
2	2	1	3.6000	.200				5.4444	.046
								5.4000	.051
2	2	2	4.5714	.067				4.5111	.098
			3.7143	.200				4.4444	.102
3	1	1	3.2000	.300	4	3	3	6.7455	.010
								6.7091	.013
3	2	1	4.2857	.100				5.7909	.046
			3.8571	.133				5.7273	.050
								4.7091	.092
3	2	2	5.3572	.029				4.7000	.101
			4.7143	.048					
			4.5000	.067	4	4	1	6.6667	.010
			4.4643	.105				6.1667	.022
								4.9667	.048
3	3	1	5.1429	.043				4.8667	.054
			4.5714	.100				4.1667	.082
			4.0000	.129				4.0667	.102
3	3	2	6.2500	.011	4	4	2	7.0364	.006
			5.3611	.032				6.8727	.011
			5.1389	.061				5.4545	.046
			4.5556	.100				5.2364	.052
			4.2500	.121				4.5545	.098
								4.4455	.103
3	3	3	7.2000	.004					
			6.4889	.011	4	4	3	7.1439	.010
			5.6889	.029				7.1364	.011
			5.6000	.050				5.5985	.049
			5.0667	.086				5.5758	.051
			4.6222	.100				4.5455	.099
								4.4773	.102
4	1	1	3.5714	.200					
					4	4	4	7.6538	.008
4	2	1	4.8214	.057				7.5385	.011
			4.5000	.076				5.6923	.049
			4.0179	.114				5.6538	.054
								4.6539	.097
4	2	2	6.0000	.014				4.5001	.104
			5.3333	.033					
			5.1250	.052	5	1	1	3.8571	.143
			4.4583	.100					
			4.1667	.105	5	2	1	5.2500	.036
								5.0000	.048
4	3	1	5.8333	.021				4.4500	.071
			5.2083	.050				4.2000	.095
			5.0000	.057				4.0500	.119
			4.0556	.093					
			3.8889	.129					

Sample sizes			H	p	Sample sizes			H	p
n_1	n_2	n_3			n_1	n_2	n_3		
5	2	2	6.5333	.008				5.6308	.050
			6.1333	.013				4.5487	.099
			5.1600	.034				4.5231	.103
			5.0400	.056					
			4.3733	.090	5	4	4	7.7604	.009
			4.2933	.122				7.7440	.011
								5.6571	.049
5	3	1	6.4000	.012				5.6176	.050
			4.9600	.048				4.6187	.100
			4.8711	.052				4.5527	.102
			4.0178	.095					
			3.8400	.123	5	5	1	7.3091	.009
								6.8364	.011
5	3	2	6.9091	.009				5.1273	.046
			6.8218	.010				4.9091	.053
			5.2509	.049				4.1091	.086
			5.1055	.052				4.0364	.105
			4.6509	.091					
			4.4945	.101	5	5	2	7.3385	.010
								7.2692	.010
5	3	3	7.0788	.009				5.3385	.047
			6.9818	.011				5.2462	.051
			5.6485	.049				4.6231	.097
			5.5152	.051				4.5077	.100
			4.5333	.097					
			4.4121	.109	5	5	3	7.5780	.010
								7.5429	.010
5	4	1	6.9545	.008				5.7055	.046
			6.8400	.011				5.6264	.051
			4.9855	.044				4.5451	.100
			4.8600	.056				4.5363	.102
			3.9873	.098					
			3.9600	.102	5	5	4	7.8229	.010
								7.7914	.010
5	4	2	7.2045	.009				5.6657	.049
			7.1182	.010				5.6429	.050
			5.2727	.049				4.5229	.099
			5.2682	.050				4.5200	.101
			4.5409	.098					
			4.5182	.101	5	5	5	8.0000	.009
								7.9800	.010
5	4	3	7.4449	.010				5.7800	.049
			7.3949	.011				5.6600	.051
			5.6564	.049				4.5600	.100
								4.5000	.102

Table I. Probabilities Associated with Values as Large as the Observed Values of χ_r^2 in the Friedman Test.

$k = 3$

| \multicolumn{2}{c|}{$N = 2$} | \multicolumn{2}{c|}{$N = 3$} | \multicolumn{2}{c|}{$N = 4$} | \multicolumn{2}{c}{$N = 5$} |

$N = 2$		$N = 3$		$N = 4$		$N = 5$	
χ_r^2	p	χ_r^2	p	χ_r^2	p	χ_r^2	p
0	1.000	.000	1.000	.0	1.000	.0	1.000
1	.833	.667	.944	.5	.931	.4	.954
3	.500	2.000	.528	1.5	.653	1.2	.691
4	.167	2.667	.361	2.0	.431	1.6	.522
		4.667	.194	3.5	.273	2.8	.367
		6.000	.028	4.5	.125	3.6	.182
				6.0	.069	4.8	.124
				6.5	.042	5.2	.093
				8.0	.0046	6.4	.039
						7.6	.024
						8.4	.0085
						10.0	.00077

$N = 6$		$N = 7$		$N = 8$		$N = 9$	
χ_r^2	p	χ_r^2	p	χ_r^2	p	χ_r^2	p
.00	1.000	.000	1.000	.00	1.000	.000	1.000
.33	.956	.286	.964	.25	.967	.222	.971
1.00	.740	.857	.768	.75	.794	.667	.814
1.33	.570	1.143	.620	1.00	.654	.889	.865
2.33	.430	2.000	.486	1.75	.531	1.556	.569
3.00	.252	2.571	.305	2.25	.355	2.000	.398
4.00	.184	3.429	.237	3.00	.285	2.667	.328
4.33	.142	3.714	.192	3.25	.236	2.889	.278
5.33	.072	4.571	.112	4.00	.149	3.556	.187
6.33	.052	5.429	.085	4.75	.120	4.222	.154
7.00	.029	6.000	.052	5.25	.079	4.667	.107
8.33	.012	7.143	.027	6.25	.047	5.556	.069
9.00	.0081	7.714	.021	6.75	.038	6.000	.057
9.33	.0055	8.000	.016	7.00	.030	6.222	.048
10.33	.0017	8.857	.0084	7.75	.018	6.889	.031
12.00	.00013	10.286	.0036	9.00	.0099	8.000	.019
		10.571	.0027	9.25	.0080	8.222	.016
		11.143	.0012	9.75	.0048	8.667	.010
		12.286	.00032	10.75	.0024	9.556	.0060
		14.000	.000021	12.00	.0011	10.667	.0035
				12.25	.00086	10.889	.0029
				13.00	.00026	11.556	.0013
				14.25	.000061	12.667	.00066
				16.00	.0000036	13.556	.00035
						14.000	.00020
						14.222	.000097
						14.889	.000054
						16.222	.000011
						18.000	.0000006

k=4

| \multicolumn{2}{c|}{$N=2$} | \multicolumn{2}{c|}{$N=3$} | \multicolumn{4}{c}{$N=4$} |

x_r^2	p	x_r^2	p	x_r^2	p	x_r^2	p
.0	1.000	.2	1.000	.0	1.000	5.7	.141
.6	.958	.6	.958	.3	.992	6.0	.105
1.2	.834	1.0	.910	.6	.928	6.3	.094
1.8	.792	1.8	.727	.9	.900	6.6	.077
2.4	.625	2.2	.608	1.2	.800	6.9	.068
3.0	.542	2.6	.524	1.5	.754	7.2	.054
3.6	.458	3.4	.446	1.8	.677	7.5	.052
4.2	.375	3.8	.342	2.1	.649	7.8	.036
4.8	.208	4.2	.300	2.4	.524	8.1	.033
5.4	.167	5.0	.207	2.7	.508	8.4	.019
6.0	.042	5.4	.175	3.0	.432	8.7	.014
		5.8	.148	3.3	.389	9.3	.012
		6.6	.075	3.6	.355	9.6	.0069
		7.0	.054	3.9	.324	9.9	.0062
		7.4	.033	4.5	.242	10.2	.0027
		8.2	.017	4.8	.200	10.8	.0016
		9.0	.0017	5.1	.190	11.1	.00094
				5.4	.158	12.0	.000072

Table J. F Distribution: 5% (Roman Type) and 10% (Boldface Type) Points for the Distribution of F.

Within group df	Between group df																					
	1	2	3	4	5	6	7	8	9	10	11	12	14	16	20	24	30	40	50	75	100	200	500	∞
1	161 4052	200 4999	216 5403	225 5625	230 5764	234 5859	237 5928	239 5981	241 6022	242 6056	243 6082	244 6106	245 6142	246 6169	248 6208	249 6234	250 6258	251 6286	252 6302	253 6323	253 6334	254 6352	254 6361	254 6366
2	18.51 98.49	19.00 99.01	19.16 99.17	19.25 99.25	19.30 99.30	19.33 99.33	19.36 99.34	19.37 99.36	19.38 99.38	19.39 99.40	19.40 99.41	19.41 99.42	19.42 99.43	19.43 99.44	19.44 99.45	19.45 99.46	19.46 99.47	19.47 99.48	19.47 99.48	19.48 99.49	19.49 99.49	19.49 99.49	19.50 99.50	19.50 99.50
3	10.13 34.12	9.55 30.81	9.28 29.46	9.12 28.71	9.01 28.24	8.94 27.91	8.88 27.67	8.84 27.49	8.81 27.34	8.78 27.23	8.76 27.13	8.74 27.05	8.71 26.92	8.69 26.83	8.66 26.69	8.64 26.60	8.62 26.50	8.60 26.41	8.58 26.30	8.57 26.27	8.56 26.23	8.54 26.18	8.54 26.14	8.53 26.12
4	7.71 21.20	6.94 18.00	6.59 16.69	6.39 15.98	6.26 15.52	6.16 15.21	6.09 14.98	6.04 14.80	6.00 14.66	5.96 14.54	5.93 14.45	5.91 14.37	5.87 14.24	5.84 14.15	5.80 14.02	5.77 13.93	5.74 13.83	5.71 13.74	5.70 13.69	5.68 13.61	5.66 13.57	5.65 13.52	5.64 13.48	5.63 13.46
5	6.61 16.26	5.79 13.27	5.41 12.06	5.19 11.39	5.05 10.97	4.95 10.67	4.88 10.45	4.82 10.27	4.78 10.15	4.74 10.05	4.70 9.96	4.68 9.89	4.64 9.77	4.60 9.68	4.56 9.55	4.53 9.47	4.50 9.38	4.46 9.29	4.44 9.24	4.42 9.17	4.40 9.13	4.38 9.07	4.37 9.04	4.36 9.02
6	5.99 13.74	5.14 10.92	4.76 9.78	4.53 9.15	4.39 8.75	4.28 8.47	4.21 8.26	4.15 8.10	4.10 7.98	4.06 7.87	4.03 7.79	4.00 7.72	3.96 7.60	3.92 7.52	3.87 7.39	3.84 7.31	3.81 7.23	3.77 7.14	3.75 7.09	3.72 7.02	3.71 6.99	3.69 6.94	3.68 6.90	3.67 6.88
7	5.59 12.25	4.74 9.55	4.35 8.45	4.12 7.85	3.97 7.46	3.87 7.19	3.79 7.00	3.73 6.84	3.68 6.71	3.63 6.62	3.60 6.54	3.57 6.47	3.52 6.35	3.49 6.27	3.44 6.15	3.41 6.07	3.38 5.98	3.34 5.90	3.32 5.85	3.29 5.78	3.28 5.75	3.25 5.70	3.24 5.67	3.23 5.65
8	5.32 11.26	4.46 8.65	4.07 7.59	3.84 7.01	3.69 6.63	3.58 6.37	3.50 6.19	3.44 6.03	3.39 5.91	3.34 5.82	3.31 5.74	3.28 5.67	3.23 5.56	3.20 5.48	3.15 5.36	3.12 5.28	3.08 5.20	3.05 5.11	3.03 5.06	3.00 5.00	2.98 4.96	2.96 4.91	2.94 4.88	2.93 4.86
9	5.12 10.56	4.26 8.02	3.86 6.99	3.63 6.42	3.48 6.06	3.37 5.80	3.29 5.62	3.23 5.47	3.18 5.35	3.13 5.26	3.10 5.18	3.07 5.11	3.02 5.00	2.98 4.92	2.93 4.80	2.90 4.73	2.86 4.64	2.82 4.56	2.80 4.51	2.77 4.45	2.76 4.41	2.73 4.36	2.72 4.33	2.71 4.31

10	4.96 / 10.04	4.10 / 7.56	3.71 / 6.55	3.48 / 5.99	3.33 / 5.64	3.22 / 5.39	3.14 / 5.21	3.07 / 5.06	3.02 / 4.95	2.97 / 4.85	2.94 / 4.78	2.91 / 4.71	2.86 / 4.60	2.82 / 4.52	2.77 / 4.41	2.74 / 4.33	2.70 / 4.25	2.67 / 4.17	2.64 / 4.12	2.61 / 4.05	2.59 / 4.01	2.56 / 3.96	2.55 / 3.93	2.54 / 3.91
11	4.84 / 9.65	3.98 / 7.20	3.59 / 6.22	3.36 / 5.67	3.20 / 5.32	3.09 / 5.07	3.01 / 4.88	2.95 / 4.74	2.90 / 4.63	2.86 / 4.54	2.82 / 4.46	2.79 / 4.40	2.74 / 4.29	2.70 / 4.21	2.65 / 4.10	2.61 / 4.02	2.57 / 3.94	2.53 / 3.86	2.50 / 3.80	2.47 / 3.74	2.45 / 3.70	2.42 / 3.66	2.41 / 3.62	2.40 / 3.60
12	4.75 / 9.33	3.88 / 6.93	3.49 / 5.95	3.26 / 5.41	3.11 / 5.06	3.00 / 4.82	2.92 / 4.65	2.85 / 4.50	2.80 / 4.39	2.76 / 4.30	2.72 / 4.22	2.69 / 4.16	2.64 / 4.05	2.60 / 3.98	2.54 / 3.86	2.50 / 3.78	2.46 / 3.70	2.42 / 3.61	2.40 / 3.56	2.36 / 3.49	2.35 / 3.46	2.32 / 3.41	2.31 / 3.38	2.30 / 3.36
13	4.67 / 9.07	3.80 / 6.70	3.41 / 5.74	3.18 / 5.20	3.02 / 4.86	2.92 / 4.62	2.84 / 4.44	2.77 / 4.30	2.72 / 4.19	2.67 / 4.10	2.63 / 4.02	2.60 / 3.96	2.55 / 3.85	2.51 / 3.78	2.46 / 3.67	2.42 / 3.59	2.38 / 3.51	2.34 / 3.42	2.32 / 3.37	2.28 / 3.30	2.26 / 3.27	2.24 / 3.21	2.22 / 3.18	2.21 / 3.16
14	4.60 / 8.86	3.74 / 6.51	3.34 / 5.56	3.11 / 5.03	2.96 / 4.69	2.85 / 4.46	2.77 / 4.28	2.70 / 4.14	2.65 / 4.03	2.60 / 3.94	2.56 / 3.86	2.53 / 3.80	2.48 / 3.70	2.44 / 3.62	2.39 / 3.51	2.35 / 3.43	2.31 / 3.34	2.27 / 3.26	2.24 / 3.21	2.21 / 3.14	2.19 / 3.11	2.16 / 3.06	2.14 / 3.02	2.13 / 3.00
15	4.54 / 8.68	3.68 / 6.36	3.29 / 5.42	3.06 / 4.89	2.90 / 4.56	2.79 / 4.32	2.70 / 4.14	2.64 / 4.00	2.59 / 3.89	2.55 / 3.80	2.51 / 3.73	2.48 / 3.67	2.43 / 3.56	2.39 / 3.48	2.33 / 3.36	2.29 / 3.29	2.25 / 3.20	2.21 / 3.14	2.18 / 3.07	2.15 / 3.00	2.12 / 2.97	2.10 / 2.92	2.08 / 2.89	2.07 / 2.87
16	4.49 / 8.53	3.63 / 6.23	3.24 / 5.29	3.01 / 4.77	2.85 / 4.44	2.74 / 4.20	2.66 / 4.03	2.59 / 3.89	2.54 / 3.78	2.49 / 3.69	2.45 / 3.61	2.42 / 3.55	2.37 / 3.45	2.33 / 3.37	2.28 / 3.25	2.24 / 3.18	2.20 / 3.10	2.16 / 3.01	2.13 / 2.96	2.09 / 2.89	2.07 / 2.86	2.04 / 2.80	2.02 / 2.77	2.01 / 2.75
17	4.45 / 8.40	3.59 / 6.11	3.20 / 5.18	2.96 / 4.67	2.81 / 4.34	2.70 / 4.10	2.62 / 3.93	2.55 / 3.79	2.50 / 3.68	2.45 / 3.59	2.41 / 3.51	2.38 / 3.45	2.33 / 3.35	2.29 / 3.27	2.23 / 3.16	2.19 / 3.08	2.15 / 3.00	2.11 / 2.92	2.08 / 2.86	2.04 / 2.79	2.02 / 2.76	1.99 / 2.70	1.97 / 2.67	1.96 / 2.65
18	4.41 / 8.28	3.55 / 6.01	3.16 / 5.09	2.93 / 4.58	2.77 / 4.25	2.66 / 4.01	2.58 / 3.85	2.51 / 3.71	2.46 / 3.60	2.41 / 3.51	2.37 / 3.44	2.34 / 3.37	2.29 / 3.27	2.25 / 3.19	2.19 / 3.07	2.15 / 3.00	2.11 / 2.91	2.07 / 2.83	2.04 / 2.78	2.00 / 2.71	1.98 / 2.68	1.95 / 2.62	1.93 / 2.59	1.92 / 2.57
19	4.38 / 8.18	3.52 / 5.93	3.13 / 5.01	2.90 / 4.50	2.74 / 4.17	2.63 / 3.94	2.55 / 3.77	2.48 / 3.63	2.43 / 3.52	2.38 / 3.43	2.34 / 3.36	2.31 / 3.30	2.26 / 3.19	2.21 / 3.12	2.15 / 3.00	2.11 / 2.92	2.07 / 2.84	2.02 / 2.76	2.00 / 2.70	1.96 / 2.63	1.94 / 2.60	1.91 / 2.54	1.90 / 2.51	1.88 / 2.49
20	4.35 / 8.10	3.49 / 5.85	3.10 / 4.94	2.87 / 4.43	2.71 / 4.10	2.60 / 3.87	2.52 / 3.71	2.45 / 3.56	2.40 / 3.45	2.35 / 3.37	2.32 / 3.31	2.28 / 3.23	2.23 / 3.13	2.18 / 3.05	2.12 / 2.94	2.08 / 2.86	2.04 / 2.77	1.99 / 2.69	1.96 / 2.63	1.92 / 2.56	1.90 / 2.53	1.87 / 2.47	1.85 / 2.44	1.84 / 2.42
21	4.32 / 8.02	3.47 / 5.78	3.07 / 4.87	2.84 / 4.37	2.68 / 4.04	2.57 / 3.81	2.49 / 3.65	2.42 / 3.51	2.37 / 3.40	2.32 / 3.31	2.28 / 3.24	2.25 / 3.17	2.20 / 3.07	2.15 / 2.99	2.09 / 2.88	2.05 / 2.80	2.00 / 2.72	1.96 / 2.63	1.93 / 2.58	1.89 / 2.51	1.87 / 2.47	1.84 / 2.42	1.82 / 2.38	1.81 / 2.36
22	4.30 / 7.94	3.44 / 5.72	3.05 / 4.82	2.82 / 4.31	2.66 / 3.99	2.55 / 3.76	2.47 / 3.59	2.40 / 3.45	2.35 / 3.35	2.30 / 3.26	2.26 / 3.18	2.23 / 3.12	2.18 / 3.02	2.13 / 2.94	2.07 / 2.83	2.03 / 2.75	1.98 / 2.67	1.93 / 2.58	1.91 / 2.53	1.87 / 2.46	1.84 / 2.42	1.81 / 2.37	1.80 / 2.33	1.78 / 2.31
23	4.28 / 7.88	3.42 / 5.66	3.03 / 4.76	2.80 / 4.26	2.64 / 3.94	2.53 / 3.71	2.45 / 3.54	2.38 / 3.41	2.32 / 3.30	2.28 / 3.21	2.24 / 3.14	2.20 / 3.07	2.14 / 2.97	2.10 / 2.89	2.04 / 2.78	2.00 / 2.70	1.96 / 2.62	1.91 / 2.53	1.88 / 2.48	1.84 / 2.41	1.82 / 2.37	1.79 / 2.32	1.77 / 2.28	1.76 / 2.26
24	4.26 / 7.82	3.40 / 5.61	3.01 / 4.72	2.78 / 4.22	2.62 / 3.90	2.51 / 3.67	2.43 / 3.50	2.36 / 3.36	2.30 / 3.25	2.26 / 3.17	2.22 / 3.09	2.18 / 3.03	2.13 / 2.93	2.09 / 2.85	2.02 / 2.74	1.98 / 2.66	1.94 / 2.58	1.89 / 2.49	1.86 / 2.44	1.82 / 2.36	1.80 / 2.33	1.76 / 2.27	1.74 / 2.23	1.73 / 2.21
25	4.24 / 7.77	3.38 / 5.57	2.99 / 4.68	2.76 / 4.18	2.60 / 3.86	2.49 / 3.63	2.41 / 3.46	2.34 / 3.32	2.28 / 3.21	2.24 / 3.13	2.20 / 3.05	2.16 / 2.99	2.11 / 2.89	2.06 / 2.81	2.00 / 2.70	1.96 / 2.62	1.92 / 2.54	1.87 / 2.45	1.84 / 2.40	1.80 / 2.32	1.77 / 2.29	1.74 / 2.23	1.72 / 2.19	1.71 / 2.17

Within group df	\multicolumn{19}{c}{Between group df}
	1	2	3	4	5	6	7	8	9	10	11	12	14	16	20	24	30	40	50	75	100	200	500	∞
26	4.22 7.72	3.37 5.53	2.89 4.64	2.74 4.14	2.59 3.82	2.47 3.59	2.39 3.42	2.32 3.29	2.27 3.17	2.22 3.09	2.18 3.02	2.15 2.96	2.10 2.86	2.05 2.77	1.99 2.66	1.95 2.58	1.90 2.50	1.85 2.41	1.82 2.36	1.78 2.28	1.76 2.25	1.72 2.19	1.70 2.15	1.69 2.13
27	4.21 7.68	3.35 5.49	2.96 4.60	2.73 4.11	2.57 3.79	2.46 3.56	2.37 3.39	2.30 3.26	2.25 3.14	2.20 3.06	2.16 2.98	2.13 2.93	2.08 2.83	2.03 2.74	1.97 2.63	1.93 2.55	1.88 2.47	1.84 2.38	1.80 2.33	1.76 2.25	1.74 2.21	1.71 2.16	1.68 2.12	1.67 2.10
28	4.20 7.64	3.34 5.45	2.95 4.57	2.71 4.07	2.56 3.76	2.44 3.53	2.36 3.36	2.29 3.23	2.24 3.11	2.19 3.03	2.15 2.95	2.12 2.90	2.06 2.80	2.02 2.71	1.96 2.60	1.91 2.52	1.87 2.44	1.81 2.35	1.78 2.30	1.75 2.22	1.72 2.18	1.69 2.13	1.67 2.09	1.65 2.06
29	4.18 7.60	3.33 5.52	2.93 4.54	2.70 4.04	2.54 3.73	2.43 3.50	2.35 3.33	2.28 3.20	2.22 3.08	2.18 3.00	2.14 2.92	2.10 2.87	2.05 2.77	2.00 2.68	1.94 2.57	1.90 2.49	1.85 2.41	1.80 2.32	1.77 2.27	1.73 2.19	1.71 2.15	1.68 2.10	1.65 2.06	1.64 2.03
30	4.17 7.56	3.32 5.39	2.92 4.51	2.69 4.02	2.53 3.70	2.42 3.47	2.34 3.30	2.27 3.17	2.21 3.06	2.16 2.98	2.12 2.90	2.09 2.84	2.04 2.74	1.99 2.66	1.93 2.55	1.89 2.47	1.84 2.38	1.79 2.29	1.76 2.24	1.72 2.16	1.69 2.13	1.66 2.07	1.64 2.03	1.62 2.01
32	4.15 7.50	3.30 5.34	2.90 4.46	2.67 3.97	2.51 3.66	2.40 3.42	2.32 3.25	2.25 3.12	2.19 3.01	2.14 2.94	2.10 2.86	2.07 2.80	2.02 2.70	1.97 2.62	1.91 2.51	1.86 2.42	1.82 2.34	1.76 2.25	1.74 2.20	1.69 2.12	1.67 2.08	1.64 2.02	1.61 1.98	1.59 1.96
34	4.13 7.44	3.28 5.29	2.88 4.42	2.65 3.93	2.49 3.61	2.38 3.38	2.30 3.21	2.23 3.08	2.17 2.97	2.12 2.89	2.08 2.82	2.05 2.76	2.00 2.66	1.95 2.58	1.89 2.47	1.84 2.38	1.80 2.30	1.74 2.21	1.71 2.15	1.67 2.08	1.64 2.04	1.61 1.98	1.59 1.94	1.57 1.91
36	4.11 7.39	3.26 5.25	2.86 4.38	2.63 3.89	2.48 3.58	2.36 3.35	2.28 3.18	2.21 3.04	2.15 2.94	2.10 2.86	2.06 2.78	2.03 2.72	1.98 2.62	1.93 2.54	1.87 2.43	1.82 2.35	1.78 2.26	1.72 2.17	1.69 2.12	1.65 2.04	1.62 2.00	1.59 1.94	1.56 1.90	1.55 1.87
38	4.10 7.35	3.25 5.21	2.85 4.34	2.62 3.86	2.46 3.54	2.35 3.32	2.26 3.15	2.19 3.02	2.14 2.91	2.09 2.82	2.05 2.75	2.02 2.69	1.96 2.59	1.92 2.51	1.85 2.40	1.80 2.32	1.76 2.22	1.71 2.14	1.67 2.08	1.63 2.00	1.60 1.97	1.57 1.90	1.54 1.86	1.53 1.84
40	4.08 7.31	3.23 5.18	2.84 4.31	2.61 3.83	2.45 3.51	2.34 3.29	2.25 3.12	2.18 2.99	2.12 2.88	2.07 2.80	2.04 2.73	2.00 2.66	1.95 2.56	1.90 2.49	1.84 2.37	1.79 2.29	1.74 2.20	1.69 2.11	1.66 2.05	1.61 1.97	1.59 1.94	1.55 1.88	1.53 1.84	1.51 1.81
42	4.07 7.27	3.22 5.15	2.83 4.29	2.59 3.80	2.44 3.49	2.32 3.26	2.24 3.10	2.17 2.96	2.11 2.86	2.06 2.77	2.02 2.70	1.99 2.64	1.94 2.54	1.89 2.46	1.82 2.35	1.78 2.26	1.73 2.17	1.68 2.08	1.64 2.02	1.60 1.94	1.57 1.91	1.54 1.85	1.51 1.80	1.49 1.78
44	4.06 7.24	3.21 5.12	2.82 4.26	2.58 3.78	2.43 3.46	2.31 3.24	2.23 3.07	2.16 2.94	2.10 2.84	2.05 2.75	2.01 2.68	1.98 2.62	1.92 2.52	1.88 2.44	1.81 2.32	1.76 2.24	1.72 2.15	1.66 2.06	1.63 2.00	1.58 1.92	1.56 1.88	1.52 1.82	1.50 1.78	1.48 1.75
46	4.05 7.21	3.20 5.10	2.81 4.24	2.57 3.76	2.42 3.44	2.30 3.22	2.22 3.05	2.14 2.92	2.09 2.82	2.04 2.73	2.00 2.66	1.97 2.60	1.91 2.50	1.87 2.42	1.80 2.30	1.75 2.22	1.71 2.13	1.65 2.04	1.62 1.98	1.57 1.90	1.54 1.86	1.51 1.80	1.48 1.76	1.46 1.72
48	4.04 7.19	3.19 5.08	2.80 4.22	2.56 3.74	2.41 3.42	2.30 3.20	2.21 3.04	2.14 2.90	2.08 2.80	2.03 2.71	1.99 2.64	1.96 2.58	1.90 2.48	1.86 2.40	1.79 2.28	1.74 2.20	1.70 2.11	1.64 2.02	1.61 1.96	1.56 1.88	1.53 1.84	1.50 1.78	1.47 1.73	1.45 1.70

df																								
50	4.03 / 7.17	3.18 / 5.06	2.79 / 4.20	2.56 / 3.72	2.40 / 3.41	2.29 / 3.18	2.20 / 3.02	2.13 / 2.88	2.07 / 2.78	2.02 / 2.70	1.98 / 2.62	1.95 / 2.56	1.90 / 2.46	1.85 / 2.39	1.78 / 2.26	1.74 / 2.18	1.69 / 2.10	1.63 / 2.00	1.60 / 1.94	1.55 / 1.86	1.52 / 1.82	1.48 / 1.76	1.46 / 1.71	1.44 / 1.68
55	4.02 / 7.12	3.17 / 5.01	2.78 / 4.16	2.54 / 3.68	2.38 / 3.37	2.27 / 3.15	2.18 / 2.98	2.11 / 2.85	2.05 / 2.75	2.00 / 2.66	1.97 / 2.59	1.93 / 2.53	1.88 / 2.43	1.83 / 2.35	1.76 / 2.23	1.72 / 2.15	1.67 / 2.06	1.61 / 1.96	1.58 / 1.90	1.52 / 1.82	1.50 / 1.78	1.46 / 1.71	1.43 / 1.66	1.41 / 1.64
60	4.00 / 7.08	3.15 / 4.98	2.76 / 4.13	2.52 / 3.65	2.37 / 3.34	2.25 / 3.12	2.17 / 2.95	2.10 / 2.82	2.04 / 2.72	1.99 / 2.63	1.95 / 2.56	1.92 / 2.50	1.86 / 2.40	1.81 / 2.32	1.75 / 2.20	1.70 / 2.12	1.65 / 2.03	1.59 / 1.93	1.56 / 1.87	1.50 / 1.79	1.48 / 1.74	1.44 / 1.68	1.41 / 1.63	1.39 / 1.60
65	3.99 / 7.04	3.14 / 4.95	2.75 / 4.10	2.51 / 3.62	2.36 / 3.31	2.24 / 3.09	2.15 / 2.93	2.08 / 2.79	2.02 / 2.70	1.98 / 2.61	1.94 / 2.54	1.90 / 2.47	1.85 / 2.37	1.80 / 2.30	1.73 / 2.18	1.68 / 2.09	1.63 / 2.00	1.57 / 1.90	1.54 / 1.84	1.49 / 1.76	1.46 / 1.71	1.42 / 1.64	1.39 / 1.60	1.37 / 1.56
70	3.98 / 7.01	3.13 / 4.92	2.74 / 4.08	2.50 / 3.60	2.35 / 3.29	2.32 / 3.07	2.14 / 2.91	2.07 / 2.77	2.01 / 2.67	1.97 / 2.59	1.93 / 2.51	1.89 / 2.45	1.84 / 2.35	1.79 / 2.28	1.72 / 2.15	1.67 / 2.07	1.62 / 1.98	1.56 / 1.88	1.53 / 1.82	1.47 / 1.74	1.45 / 1.69	1.40 / 1.63	1.37 / 1.56	1.35 / 1.53
80	3.96 / 6.96	3.11 / 4.88	2.72 / 4.04	2.48 / 3.56	2.33 / 3.25	2.21 / 3.04	2.12 / 2.87	2.05 / 2.74	1.99 / 2.64	1.95 / 2.55	1.91 / 2.48	1.88 / 2.41	1.82 / 2.32	1.77 / 2.24	1.70 / 2.11	1.65 / 2.03	1.60 / 1.94	1.54 / 1.84	1.51 / 1.78	1.45 / 1.70	1.42 / 1.65	1.38 / 1.57	1.35 / 1.52	1.32 / 1.49
100	3.94 / 6.90	3.09 / 4.82	2.70 / 3.98	2.46 / 3.51	2.30 / 3.20	2.19 / 2.99	2.10 / 2.82	2.03 / 2.69	1.97 / 2.59	1.92 / 2.51	1.88 / 2.43	1.85 / 2.36	1.79 / 2.26	1.75 / 2.19	1.68 / 2.06	1.63 / 1.98	1.57 / 1.89	1.51 / 1.79	1.48 / 1.73	1.42 / 1.64	1.39 / 1.59	1.34 / 1.51	1.30 / 1.46	1.28 / 1.43
125	3.92 / 6.84	3.07 / 4.78	2.68 / 3.94	2.44 / 3.47	2.29 / 3.17	2.17 / 2.95	2.08 / 2.79	2.01 / 2.65	1.95 / 2.56	1.90 / 2.47	1.86 / 2.40	1.83 / 2.33	1.77 / 2.23	1.72 / 2.15	1.65 / 2.03	1.60 / 1.94	1.55 / 1.85	1.49 / 1.75	1.45 / 1.68	1.39 / 1.59	1.36 / 1.54	1.31 / 1.46	1.27 / 1.40	1.25 / 1.37
150	3.91 / 6.81	3.06 / 4.75	2.67 / 3.91	2.43 / 3.44	2.27 / 3.13	2.16 / 2.92	2.07 / 2.76	2.00 / 2.62	1.94 / 2.53	1.89 / 2.44	1.85 / 2.37	1.82 / 2.30	1.76 / 2.20	1.71 / 2.12	1.64 / 2.00	1.59 / 1.91	1.54 / 1.83	1.47 / 1.72	1.44 / 1.66	1.37 / 1.56	1.34 / 1.51	1.29 / 1.43	1.25 / 1.37	1.22 / 1.33
200	3.89 / 6.76	3.04 / 4.71	2.65 / 3.88	2.41 / 3.41	2.26 / 3.11	2.14 / 2.90	2.05 / 2.73	1.98 / 2.60	1.92 / 2.51	1.87 / 2.41	1.83 / 2.34	1.80 / 2.28	1.74 / 2.17	1.69 / 2.09	1.62 / 1.97	1.57 / 1.88	1.52 / 1.79	1.45 / 1.69	1.42 / 1.62	1.35 / 1.53	1.32 / 1.48	1.26 / 1.39	1.22 / 1.33	1.19 / 1.28
400	3.86 / 6.70	3.02 / 4.66	2.62 / 3.83	2.39 / 3.36	2.23 / 3.06	2.12 / 2.85	2.03 / 2.69	1.96 / 2.55	1.90 / 2.46	1.85 / 2.37	1.81 / 2.29	1.78 / 2.23	1.72 / 2.12	1.67 / 2.04	1.60 / 1.92	1.54 / 1.84	1.49 / 1.74	1.42 / 1.64	1.38 / 1.57	1.32 / 1.47	1.28 / 1.42	1.22 / 1.32	1.16 / 1.24	1.13 / 1.19
1000	3.85 / 6.66	3.00 / 4.62	2.61 / 3.80	2.38 / 3.34	2.22 / 3.04	2.10 / 2.82	2.02 / 2.66	1.95 / 2.53	1.89 / 2.43	1.84 / 2.34	1.80 / 2.26	1.76 / 2.20	1.70 / 2.09	1.65 / 2.01	1.58 / 1.89	1.53 / 1.81	1.47 / 1.71	1.41 / 1.61	1.36 / 1.54	1.30 / 1.44	1.26 / 1.38	1.19 / 1.28	1.13 / 1.19	1.08 / 1.11
∞	3.84 / 6.64	2.99 / 4.60	2.60 / 3.78	2.37 / 3.32	2.21 / 3.02	2.09 / 2.80	2.01 / 2.64	1.94 / 2.51	1.88 / 2.41	1.83 / 2.32	1.79 / 2.24	1.75 / 2.18	1.69 / 2.07	1.64 / 1.99	1.57 / 1.87	1.52 / 1.79	1.46 / 1.69	1.40 / 1.59	1.35 / 1.52	1.28 / 1.41	1.24 / 1.36	1.17 / 1.25	1.11 / 1.15	1.00 / 1.00

Table K. Critical Values for the Spearman Rank Correlation Coefficient When N is between 4 and 10.

N	Significance level for a one-tail test	
	.05	.01
	Significance level for a two-tail test	
	.10	.02
4	1.000	
5	.900	1.000
6	.829	.943
7	.714	.893
8	.643	.833
9	.600	.783
10	.564	.746

Table L. Critical Values for the Correlation Coefficient.

	Level of significance for a one-tail test				
	.05	.025	.01	.005	.0005
	Level of significance for a two-tail test				
df	.10	.05	.02	.01	.001
1	.9877	.9969	.9995	.9999	1.0000
2	.9000	.9500	.9800	.9900	.9990
3	.8054	.8783	.9343	.9587	.9912
4	.7293	.8114	.8822	.9172	.9741
5	.6694	.7545	.8329	.8745	.9507
6	.6215	.7067	.7887	.8343	.9249
7	.5822	.6664	.7498	.7977	.8982
8	.5494	.6319	.7155	.7646	.8721
9	.5214	.6021	.6851	.7348	.8471
10	.4973	.5760	.6581	.7079	.8233
11	.4762	.5529	.6339	.6835	.8010
12	.4575	.5324	.6120	.6614	.7800
13	.4409	.5139	.5923	.6411	.7603
14	.4259	.4973	.5742	.6226	.7420
15	.4124	.4821	.5577	.6055	.7246
16	.4000	.4683	.5425	.5897	.7084
17	.3887	.4555	.5285	.5751	.6932
18	.3783	.4438	.5155	.5614	.6787
19	.3687	.4329	.5034	.5487	.6652
20	.3598	.4227	.4921	.5368	.6524
25	.3223	.3809	.4451	.4869	.5974
30	.2960	.3494	.4093	.4487	.5541
35	.2746	.3246	.3810	.4182	.5189
40	.2573	.3044	.3578	.3932	.4896
45	.2428	.2875	.3384	.3721	.4648
50	.2306	.2732	.3218	.3541	.4433
60	.2108	.2500	.2948	.3248	.4078
70	.1954	.2319	.2737	.3017	.3799
80	.1829	.2172	.2565	.2830	.3568
90	.1726	.2050	.2422	.2673	.3375
100	.1638	.1946	.2301	.2540	.3211

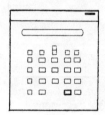

Appendix B
Statistical Formulas

Mean (Chapter 2)

$$\overline{X} = \frac{\Sigma X}{N}$$

Standard Deviation (Chapter 2)

$$s = \frac{1}{N} \sqrt{(N)\Sigma X^2 - (\Sigma X)^2}$$

Standard Error of the Mean (Chapter 2)

$$S.E.M. = \frac{s}{\sqrt{N}}$$

Chi Square Test (Chapter 5)

$$\chi^2 = \Sigma \frac{(O-E)^2}{E}$$

McNemar Test (Chapter 5)

$$\chi^2 = \frac{(|A-D|-1)^2}{A+D}$$

Cochran Q Test (Chapter 5)

$$Q = \frac{(k-1)(k\Sigma C^2 - T^2)}{kT - \Sigma R^2}$$

Kolmogorov-Smirnov Test (Chapter 6)

$$D = \frac{LD}{N}$$

Mann-Whitney U (Chapter 6)

$$U_1 = n_1 n_2 + \frac{n_1(n_1+1)}{2} - \Sigma R_1$$

Mann-Whitney U (Chapter 6)

$$U_2 = n_1 n_2 - U_1$$

Mann-Whitney U (when n_2 is larger than 20) (Chapter 6)

$$z = \frac{U + \frac{1}{2} - \frac{n_1 n_2}{2}}{\sqrt{\frac{n_1 n_2 (n_1 + n_2 + 1)}{12}}}$$

Wilcoxon Test (Chapter 6)

T-Smaller Value for either ΣR^+ or ΣR^-

Wilcoxon Test (sample size larger than 25) (Chapter 6)

$$z = \frac{T - \frac{N(N+1)}{4}}{\sqrt{\frac{N(N+1)(2N+1)}{24}}}$$

Kruskal-Wallis Test (Chapter 6)

$$H = \frac{12}{N(N+1)} \left[\frac{(\Sigma R_1)^2}{n_1} + \frac{(\Sigma R_2)^2}{n_2} + \frac{(\Sigma R_3)^2}{n_3} + \cdots + \frac{(\Sigma R_k)^2}{n_k} \right] - 3(N+1)$$

Friedman Test (Chapter 6)

$$\chi r^2 = \frac{12}{Nk(k+1)} \ [\Sigma R_i^2] - 3N(k+1)$$

t-Test (I) (Chapter 7)

$$t = \frac{\overline{X} - \mu}{\sqrt{\frac{\Sigma X^2 - \frac{(\Sigma X)^2}{N}}{N(N-1)}}}$$

t-Test (II) (Chapter 7)

$$\frac{\overline{X}_1 - \overline{X}_2}{\sqrt{\frac{\Sigma X_1^2 - \frac{(\Sigma X_1)^2}{N_1} + \Sigma X_2^2 - \frac{(\Sigma X_2)^2}{N_2}}{N_1 + N_2 - 2} \cdot \left(\frac{N_1 + N_2}{N_1 \cdot N_2} \right)}}$$

t-Test (III) (Chapter 7)

$$t = \frac{\overline{D}}{\sqrt{\frac{\Sigma D^2 - \frac{(\Sigma D)^2}{N}}{N(N-1)}}}$$

One-Way Analysis of Variance Randomized Groups (Chapter 7)

F-Ratio
$$F = \frac{MS_B}{MS_W}$$

Scheffé Test (Chapter 7)

$$F = \frac{(\overline{X}_1 - \overline{X}_2)^2}{MS_W \left(\frac{n_1 + n_2}{n_1 n_2}\right)(k-1)}$$

One-Way Analysis of Variance Randomized Blocks (Chapter 7)

F-Ratio
$$F = \frac{MS_c}{MS_e}$$

Contingency Coefficient (Chapter 8)

$$C = \sqrt{\frac{\chi^2}{\chi^2 + N}}$$

Spearman Rho Coefficient (Chapter 8)

$$\rho = 1 - \frac{6\Sigma D^2}{N(N^2 - 1)}$$

Spearman when N is 30 or More (Chapter 8)

$$z = \rho \sqrt{N-1}$$

Pearson r Coefficient (Chapter 8)

$$r = \frac{N\Sigma XY - (\Sigma X)(\Sigma Y)}{\sqrt{[N\Sigma X^2 - (\Sigma X)^2][N\Sigma Y^2 - (\Sigma Y)^2]}}$$

Pearson when N is 30 or More (Chapter 8)
$$z = r\sqrt{N-1}$$

Multiple Correlation (Chapter 8)
$$R_{1.23} = \sqrt{\frac{r_{12}^2 + r_{13}^2 - 2r_{12}r_{13}r_{23}}{1 - r_{23}^2}}$$

Point-Biserial Correlation (Chapter 9)
$$r_{pb} = \frac{\overline{Y_1} - \overline{Y_0}}{\sqrt{\frac{\Sigma Y^2}{N-} - \frac{(\Sigma Y)^2}{N(N-1)}}} \sqrt{\frac{N_1 N_0}{N(N-1)}}$$

Testing for significance for the Point-Biserial Correlation (Chapter 9)

$$t = r_{pb}\sqrt{\frac{n-2}{1-r_{pb}^2}}$$

Kendall Coefficient of Concordance (W) (Chapter 9)
$$W = \frac{\Sigma\left(R_j - \frac{\Sigma R_j}{N}\right)^2}{\frac{1}{12}k^2(N^3 - N)}$$

When N is more than 7

Testing for Significance for the Kendall Coefficient of Concordance (Chapter 9)
$$\chi^2 = k(N-1)W$$

Kendall tau (Chapter 9)
$$\text{tau} = \frac{P-Q}{\frac{1}{2}N(N-1)}$$

Testing for Significance Kendall Tau (Chapter 9)
$$z = \frac{T}{\sqrt{2(2N+5)/[9N(N-1)]}}$$

Bibliography

Bradley, J.V. *Distribution-Free Statistical Tests*, Englewood Cliffs, N.J.: Prentice-Hall, 1968.

Cochran, W.G. "Some Consequences When the Assumptions Underlying the Analysis of Variance Have Not Been Met," *Biometrics* 3(1947):22-28.

Dixon, W. J., and Massey, F. J. *Introduction to Statistical Analysis*, New York: McGraw-Hill Book Co., 1957.

Downie, N.M., and Heath, R.W. *Statistical Methods* 4th ed, New York: Harper & Row, 1974.

Edwards, A.L. *Experimental Design in Psychological Research* 4th ed, New York: Holt, Rinehart & Winston, 1972.

Ferguson, G.A. *Statistical Analysis in Psychology and Education* 2nd ed, New York: McGraw-Hill Book Co., 1966.

Goodman, L.A. "Kolmogorov-Smirnov Tests for Psychological Research," *Psychological Bulletin* 51(1954):160-168.

Guilford, J.P., and Fruchter, B. *Fundamental Statistics in Psychology and Education* 5th ed, New York: McGraw-Hill Book Co., 1973.

Hoel, P. *Elementary Statistics* 4th ed, New York: John Wiley & Sons, 1976.

Hunter, W.L. *Getting the Most Out of Your Electronic Calculator*, Blue Ridge Summit, PA, TAB Books, 1974.

Lindgren, B.M. *Statistical Theory* 3rd ed, New York: Macmillan, 1976.

Lindquist, E.F. *Design and Analysis Experiment in Psychology and Education*, Boston, Houghton Mifflin, 1953.

Massey, F. J., Jr. "The Kolmogorov-Smirnov Test for Goodness of Fit," *Journal of American Statistics Association* 46(1951):68-78.

Mood, A. M. *Introduction to Theory of Statistics*, New York: McGraw-Hill Book Co. 1950.

Moses, E. L. "Non-Parametric Statistics for Psychological Research," *Psychological Bulletin*, 49(1952):122-143.

Peters, C. C., and Van Voorhis, W. R. *Statistical Procedures and Their Mathematical Bases*, New York: McGraw-Hill Book Co., 1940.

Pirie, W. R., and Hamden, M. A. "Some Revised Continuity Corrections for Discrete Data," *Biometrics* 28(1972):693-701.

Ryan, T. A. "Significance Tests for Multiple Comparison of Proportions, Variances and Other Statistics," *Psychological Bulletin* 57, 4(1960):318-328.

Scheffé, H. *The Analysis of Variance*, New York: Wiley, 1957.

Siegel, S. *Nonparametric Statistics for the Behavioral Sciences*, New York: McGraw-Hill Book Co., 1956.

Snedecor, G. W., and Cochran, W. G. *Statistical Methods* 6th ed, Ames: Iowa State College Press, 1967.

Stevens, S. S. "Measurement, Statistics, and the Schemapiric View," *Science* 161(1968):845-856.

Waler, H. M., and Lev, J. *Statistical Inference*, New York: Holt, Rinehart & Winston, 1953.

Walker, H. M. and Lev, J. *Elementary Statistical Methods* 3rd ed, New York: Holt, Rinehart & Winston, 1969.

Wilcoxon, F., and Wilcox, R. *Some Rapid Approximate Statistical Procedures*, Pearl River, New York: Lederle Laboratories, 1964, pp., 9-12.

Winer, B. J. *Statistical Principles in Experimental Design* 2nd ed, New York: McGraw-Hill Book Co., 1971.

Index

A
Alpha	31
Alternative hypothesis	31
Arithmetic symbols	3

C
Chi square (I) test	41-47
Chi square (II) test	47-53
Cluster	29
Cochran Q test	57-64
Contingency coefficient	153-157
Correlational techniques	152-201

D
Decimals	4
Degrees of freedom	31
Dependent variable	30
Descriptive statistics	27

E
Experimental study	30

F
Friedman test	92-99

H
Homoscedasticity	169

I
Independent variable	30
Inferential statistics	28
Interval tests	100-151

K
Kendall coefficient	188-194
Kendal rank correlation (tau)	195-200
Keys, calculator	20
Kolmogorov-Smirnov test	65-70
Kruskal-Wallis test	85-92

L
Level of measurement	35
Level of significance	31

M
Mann-Whitney U test	70-78
McNemar test	53-57
Mean	21
Memory	7
Memory minus	8
Memory plus	7

N
Nature of groups	38
Negative numbers	9
Nominal tests	41-64
Nonparametric tests	32
Null hypothesis	30
Number line	9

O
One-tail tests	31
Operation order	12
Ordinal tests	65-99

P

Parametric tests	32
Pearson Product-moment	162-171
Point-biserial correlation	177-188
Population	28
Positive numbers	9
Powers of numbers	6

R

Randomized block model	138-150
Randomized groups model	124-133

S

Sampling	28
Sampling distribution	31
Scattergrams	163-164
Scheffé test	133-138
Simple multiple correlation	171-175
Simple random	28
Spearman rank coefficient	157-162
Square roots	7
Squaring	6
Standard deviation	22
Standard error of the mean	24
Statistical road map	36-37
Statistical symbols	13
Stratified	29
Symbols, arithmetic	3
statistical	13

T

T-test (I)	100-105
T-test (II)	105-116
T-test (III)	116-123
Two-tail tests	31

V

Variance	32

W

Wilcoxon signed-ranks test	79-85

Z

Zero	5

519.5   S531hP
Sharp, Vicki F
  How to solve statis-
tical problems with
your pocket calculator

ML '83   0 3 4 7 8